2019: An Astronomical Year

A Reference Guide to 365 Nights of Astronomy

Richard J. Bartlett

Top Cover Image: This striking image shows ESO's Paranal Observatory in Chile soon after sunset. The last rays of the day create a spectacular orange haze as they pass through the dusty lower levels of the atmosphere, setting a perfect scene for this picture of the week.

In this long exposure image we can see star trails caused by the movement of stars across the sky as the earth rotates. These tracks look a little like dotted lines, an effect caused by combining a number of individual shots taken with short gaps in between. The crookedness at the bottom of the star trails is due to the camera moving out of place.

The path of the crescent Moon can also be seen towards the lower left of the frame as it slowly sets, appearing to sink into the Pacific Ocean. The moon is not trailed as it was taken with a series of very short exposures.

In the centre of the image a meteor is visible as a vertical streak, and the blinking lights of various planes can be seen tracking horizontally across the streaked sky.

This image was taken by ESO Photo Ambassador Gabriel Brammer soon after sunset on January 5, 2011. At the time Gabriel worked as an astronomer at the La Silla-Paranal Observatory, but is currently an astronomer supporting the Hubble Space Telescope at the Space Telescope Science Institute in Maryland, USA. This image was one of Brammer's first ever night shots of Paranal, and remains one of his favourites.

Credit: ESO/G. Brammer

Image source: http://www.eso.org/public/images/potw1503a/

Bottom Cover Image: The intricate details of the magnificent Milky Way — its dust lanes and blankets of gas — are revealed in this photograph, taken at ESO's La Silla Observatory. To the right of the image, the large reddish orb is actually the result of a Lunar Eclipse — when the Moon moves into Earth's shadow — set against a star-studded night sky.

Credit: ESO/B. Tafreshi (twanight.org)

First Edition, May 2018

Contents

Introduction

About the 2019 Edition

I made a major change to the style and format of the book last year and it seems to have been successful; or to put it another way, I didn't receive any negative comments and the online reviews were favorable.

With that in mind, I've kept the format and have only made a few small changes.

Last year, I had several pages summarizing the Moon and planets and which other stars and planets they appeared close to. I haven't included that information this year; not because I didn't feel it was useful, but rather because I didn't like the formatting and the manner in which the information was presented.

Instead I've replaced it with a table showing the AM/PM visibility of the planets with a rating for each one. So, for example, a planet at its best visibility will have a five star (*****) rating while a planet that's not visible at all will simply have a blank space.

The rating is based upon the planet's brightness, apparent diameter and elongation (angular distance) from the Sun in the sky. It's not perfect, but it provides a good, at-a-glance guide to the best times for planetary observations throughout the year.

On that same page you'll find a summary table of eclipses while on the next page you'll find a table detailing the major planetary events of the year. These are typically oppositions and close conjunctions, but there's also a transit of Mercury to look forward to.

Lastly, there's a table detailing the major meteor showers of the year. Besides the standard info (such as start and end dates etc) there are also ratings for speed and brightness. The slower and brighter the meteor, the higher the rating (since a slow and bright meteor should be easier to see.) There's also an indication of the moon phase on the date of the meteor shower's peak.

The other, smaller change involves the mini charts showing the positions of the planets for each ten day period. Last year, I had a graphic showing the position of Uranus but nothing for the Sun. This year I've flipped things around; Uranus is too slow moving and a little too faint to make the chart useful and I thought depicting the position of the Sun would have a greater benefit. (I couldn't include a Uranus chart for formatting reasons. If you'll pardon the pun, I simply didn't have the space.)

For the sake of clarity, I've also changed the style of the Sun/planet position markers on the charts.

As always, the charts were compiled using Wolfgang Zima's excellent *Mobile Observatory* app for Android devices. You can download it from the Google Play store or check it out at http://zima.co

The images of the planets were also created using the app. Each image shows the view you might get through a very high magnification eyepiece and spans the equivalent of one arc second. That way you can get an idea of how much larger or smaller a planet might appear in comparison to another.

The last change involves the book as a whole. In previous years, I've produced two separate guides, one for the United Kingdom and one for the United States. This has involved some extra work so this year I've produced one, universal guide. The information is applicable to both sides of the Atlantic but the star charts are better suited to North American skies. (They're still perfectly usable in the U.K. but stars and constellations over the southern horizon will appear lower and, conversely, stars and constellations over the northern horizon will appear higher.)

You can download PDF charts and observing lists specifically for the U.K. at the following URL: https://tinyurl.com/ukstarcharts

I'm always looking for ways to improve the book and am open to comments and suggestions. With that in mind, please feel free to email me at astronomywriter@gmail.com if there's anything you'd like to see included.

About the Author

Photo by my son, James Bartlett

I've had an interest in astronomy since I was six and although my interest has waxed and waned like the Moon, I've always felt compelled to stop and stare at the stars.

In the late 90's, I discovered the booming frontier of the internet, and like a settler in the Midwest, I quickly staked my claim on it. I started to build a (now-defunct) website called *StarLore.* It was designed to be an online resource for amateur astronomers who wanted to know more about the constellations - and all the stars and deep sky objects to be found within them. It was quite an undertaking.

After the website was featured in the February 2001 edition of *Sky & Telescope* magazine, I began reviewing astronomical websites and software for their rival, *Astronomy*. This was something of a dream come true; I'd been reading the magazine since I was a kid and now my name was regularly appearing in it.

Unfortunately, a financial downturn forced my monthly column to be cut after a few years but I'll always be grateful for the chance to write for the world's best-selling astronomy magazine.

I emigrated from England to the United States in 2004 and spent three years under relatively clear, dark skies in Oklahoma. I then relocated to Kentucky in 2008 and then California in 2013. I now live in the suburbs of Los Angeles; not the most ideal location for astronomy, but there are still a number of naked eye events that are easily visible on any given night.

The Author Online

Amazon US: http://tinyurl.com/rjbamazon-us

Amazon UK: http://tinyurl.com/rjbamazon-uk

Facebook: http://tinyurl.com/rjbfacebook

Twitter: http://tinyurl.com/rjbtwitter

Blog: http://tinyurl.com/astronomicalyear

Email: astronomywriter@gmail.com

AstroNews: http://tinyurl.com/astronewsus

Clear skies,

Richard J. Bartlett

May 31st, 2018

Star Charts & Observing Lists

Star Chart Tables

If observing during daylight savings time, first deduct one hour and then refer to the corresponding chart number. For example, for 10pm daylight savings time in early August, use chart 18.

	6pm	7pm	8pm	9pm	10pm	11pm
Early January	1	2	3	4	5	6
Late January	2	3	4	5	6	7
Early February	3	4	5	6	7	8
Late February	4	5	6	7	8	9
Early March	5	6	7	8	9	10
Late March	6	7	8	9	10	11
Early April	7	8	9	10	11	12
Late April	8	9	10	11	12	13
Early May	9	10	11	12	13	14
Late May	10	11	12	13	14	15
Early June	11	12	13	14	15	16
Late June	12	13	14	15	16	17
Early July	13	14	15	16	17	18
Late July	14	15	16	17	18	19
Early August	15	16	17	18	19	20
Late August	16	17	18	19	20	21
Early September	17	18	19	20	21	22
Late September	18	19	20	21	22	23
Early October	19	20	21	22	23	24
Late October	20	21	22	23	24	1
Early November	21	22	23	24	1	2
Late November	22	23	24	1	2	3
Early December	23	24	1	2	3	4
Late December	24	1	2	3	4	5

If observing during daylight savings time, first deduct one hour and then refer to the corresponding chart number. For example, for 2am daylight savings time in early July, use chart 20.

	12am	1am	2am	3am	4am	5am	6am
Early January	7	8	9	10	11	12	13
Late January	8	9	10	11	12	13	14
Early February	9	10	11	12	13	14	15
Late February	10	11	12	13	14	15	16
Early March	11	12	13	14	15	16	17
Late March	12	13	14	15	16	17	18
Early April	13	14	15	16	17	18	19
Late April	14	15	16	17	18	19	20
Early May	15	16	17	18	19	20	21
Late May	16	17	18	19	20	21	22
Early June	17	18	19	20	21	22	23
Late June	18	19	20	21	22	23	24
Early July	19	20	21	22	23	24	1
Late July	20	21	22	23	24	1	2
Early August	21	22	23	24	1	2	3
Late August	22	23	24	1	2	3	4
Early September	23	24	1	2	3	4	5
Late September	24	1	2	3	4	5	6
Early October	1	2	3	4	5	6	7
Late October	2	3	4	5	6	7	8
Early November	3	4	5	6	7	8	9
Late November	4	5	6	7	8	9	10
Early December	5	6	7	8	9	10	11
Late December	6	7	8	9	10	11	12

Chart 1

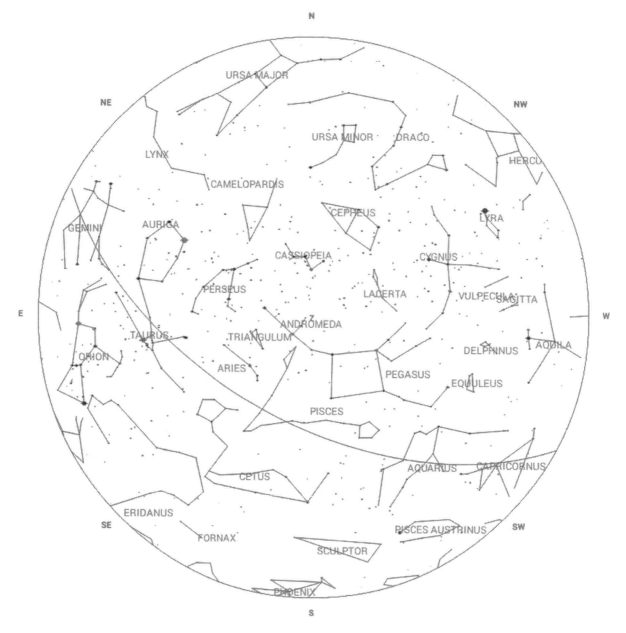

Designation	Name	Con.	Type	R.A.	Dec.	Mag	Size/Sep
Gam And	Almach	And	MS	02h 04m	+42° 20'	2.1	10"
M 31	Andromeda Galaxy	And	Gx	00h 43m	+41° 16'	4.3	156'
NGC 752	Golf Ball Cluster	And	OC	01h 58m	+37° 47'	6.6	75'
Pi And		And	MS	00h 38m	+33° 49'	4.4	36"
107 Aqr		Aqr	MS	23h 47m	-18° 35'	5.3	7"
94 Aqr		Aqr	MS	23h 19m	-13° 28'	5.2	13"
NGC 7293	Helix Nebula	Aqr	PN	22h 30m	-20° 50'	6.3	16'
Zet Aqr		Aqr	MS	22h 29m	-00° 01'	3.7	2"

Designation	Name	Con.	Type	R.A.	Dec.	Mag	Size/Sep
41 Aqr		Aqr	MS	22h 14m	-21° 04'	5.3	5"
53 Aqr		Aqr	MS	22h 27m	-16° 45'	5.6	3"
Lam Ari		Ari	MS	01h 59m	+23° 41'	4.8	37"
Gam Ari	Mesarthim	Ari	MS	01h 54m	+19° 22'	4.6	8"
30 Ari		Ari	MS	02h 37m	+24° 39'	6.5	39"
Kemble 1	Kemble's Cascade	Cam	Ast	03h 57m	+63° 04'	5.0	180'
Sig Cas		Cas	MS	23h 59m	+55° 45'	4.9	3"
NGC 457	Owl Cluster	Cas	OC	01h 20m	+58° 17'	5.1	20'
M 52		Cas	OC	23h 25m	+61° 36'	8.2	15'
Struve 163		Cas	MS	01h 51m	+64° 51'	6.5	35"
Struve 3053		Cas	MS	00h 03m	+66° 06'	5.9	15"
Eta Cas	Achird	Cas	MS	00h 50m	+57° 54'	3.6	13"
M 103		Cas	OC	01h 33m	+60° 39'	6.9	5'
NGC 281		Cas	OC	00h 53m	+56° 38'	7.4	4'
Iot Cas		Cas	MS	02h 29m	+67° 24'	4.5	7"
NGC 7789	Herschel's Spiral Cluster	Cas	OC	23h 57m	+56° 43'	7.5	25'
NGC 559		Cas	OC	01h 30m	+63° 18'	7.4	6'
NGC 659	Ying Yang Cluster	Cas	OC	01h 44m	+60° 40'	7.2	5'
NGC 663		Cas	OC	01h 46m	+61° 14'	6.4	14'
Del Cep		Cep	MS/Var	22h 29m	+58° 25'	3.5-4.4	41"
Xi Cep	Alkurhah	Cep	MS	22h 04m	+64° 38'	4.3	8"
Omi Cep		Cep	MS	23h 19m	+68° 07'	4.8	3"
Gam Cet	Kaffajidhma	Cet	MS	02h 43m	+03° 14'	3.5	3"
Omi Cet	Mira	Cet	Var	02h 19m	-02° 59'	2.0-10.1	N/A
NGC 7243		Lac	OC	22h 15m	+49° 54'	6.7	29'
8 Lac		Lac	MS	22h 36m	+39° 38'	5.7	82"
Bet Per	Algol	Per	Var	03h 08m	+40° 57'	2.1-3.4	N/A
Eps Per		Per	MS	03h 58m	+40° 01'	2.9	9"
M 34		Per	OC	02h 42m	+42° 46'	5.8	35'
NGC 869/884	Double Cluster	Per.	OC	02h 21m	+57° 08'	4.4	18'
Eta Per		Per	MS	02h 51m	+55° 54'	3.8	29"
NGC 1245		Per	OC	03h 15m	+47° 15'	7.7	10'
Melotte 20	Alpha Persei Moving Cluster	Per	OC	03h 24m	+49° 52'	2.3	300'
TX Psc		Psc	Var/CS	23h 46m	+03° 29'	4.5-5.3	N/A
Alp Psc	Alrisha	Psc	MS	02h 02m	+02° 46'	3.8	2"
Psi1 Psc		Psc	MS	01h 06m	+21° 28'	5.3	30"
Zet Psc		Psc	MS	01h 14m	+07° 35'	5.2	23"
55 Psc		Psc	MS	00h 40m	+21° 26'	5.4	6"
65 Psc		Psc	MS	00h 50m	+27° 43'	7.0	4"
M 45	Pleiades	Tau	OC	03h 47m	+24° 07'	1.5	120'
M 33	Triangulum Galaxy	Tri	Gx	01h 34m	+30° 40'	6.4	62'
Alp UMi	Polaris	UMi	MS	02h 51m	+89° 20'	2.0	18"

Chart 2

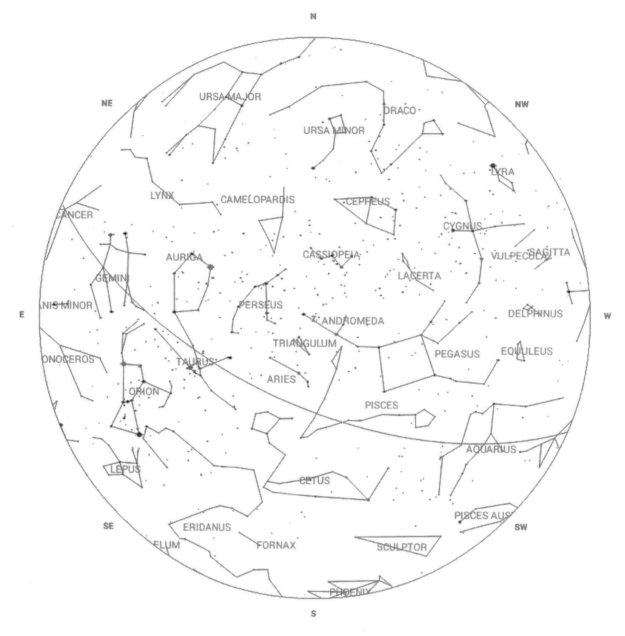

Designation	Name	Con.	Type	R.A.	Dec.	Mag	Size/Sep
Gam And	Almach	And	MS	02h 04m	+42° 20'	2.1	10"
M 31	Andromeda Galaxy	And	Gx	00h 43m	+41° 16'	4.3	156'
NGC 752	Golf Ball Cluster	And	OC	01h 58m	+37° 47'	6.6	75'
NGC 7662	Blue Snowball Nebula	And	PN	23h 26m	+42° 32'	8.6	17"
Pi And		And	MS	00h 38m	+33° 49'	4.4	36"
Lam Ari		Ari	MS	01h 59m	+23° 41'	4.8	37"
Gam Ari	Mesarthim	Ari	MS	01h 54m	+19° 22'	4.6	8"
30 Ari		Ari	MS	02h 37m	+24° 39'	6.5	39"

Designation	Name	Con.	Type	R.A.	Dec.	Mag	Size/Sep
U Cam		Cam	Var/CS	03h 42m	+62° 39'	7.0-7.5	N/A
ST Cam		Cam	Var/CS	04h 51m	+68° 10'	7.0-8.4	N/A
Kemble 1	Kemble's Cascade	Cam	Ast	03h 57m	+63° 04'	5.0	180'
NGC 1502	Jolly Roger Cluster	Cam	OC	04h 08m	+62° 20'	4.1	8'
NGC 457	Owl Cluster	Cas	OC	01h 20m	+58° 17'	5.1	20'
Iot Cas		Cas	MS	02h 29m	+67° 24'	4.5	7"
Sig Cas		Cas	MS	23h 59m	+55° 45'	4.9	3"
Struve 163		Cas	MS	01h 51m	+64° 51'	6.5	35"
Struve 3053		Cas	MS	00h 03m	+66° 06'	5.9	15"
Eta Cas	Achird	Cas	MS	00h 50m	+57° 54'	3.6	13"
M 103		Cas	OC	01h 33m	+60° 39'	6.9	5'
M 52		Cas	OC	23h 25m	+61° 36'	8.2	15'
NGC 281		Cas	OC	00h 53m	+56° 38'	7.4	4'
NGC 559		Cas	OC	01h 30m	+63° 18'	7.4	6'
NGC 659	Ying Yang Cluster	Cas	OC	01h 44m	+60° 40'	7.2	5'
NGC 663		Cas	OC	01h 46m	+61° 14'	6.4	14'
NGC 7789	Herschel's Spiral Cluster	Cas	OC	23h 57m	+56° 43'	7.5	25'
Omi Cep		Cep	MS	23h 19m	+68° 07'	4.8	3"
Gam Cet	Kaffajidhma	Cet	MS	02h 43m	+03° 14'	3.5	3"
Omi Cet	Mira	Cet	Var	02h 19m	-02° 59'	2.0-10.1	N/A
32 Eri		Eri	MS	03h 54m	-02° 57'	4.4	7"
40 Eri	Keid	Eri	MS	04h 15m	-07° 39'	4.4	83"
NGC 1528	m & m Double Cluster	Per	OC	04h 15m	+51° 13'	6.4	16'
Bet Per	Algol	Per	Var	03h 08m	+40° 57'	2.1-3.4	N/A
Eps Per		Per	MS	03h 58m	+40° 01'	2.9	9"
NGC 1245		Per	OC	03h 15m	+47° 15'	7.7	10'
M 34		Per	OC	02h 42m	+42° 46'	5.8	35'
NGC 869/884	Double Cluster	Per	OC	02h 21m	+57° 08'	4.4	18'
Eta Per		Per	MS	02h 51m	+55° 54'	3.8	29"
Melotte 20	Alpha Persei Moving Cluster	Per	OC	03h 24m	+49° 52'	2.3	300'
Alp Psc	Alrisha	Psc	MS	02h 02m	+02° 46'	3.8	2"
Psi1 Psc		Psc	MS	01h 06m	+21° 28'	5.3	30"
Zet Psc		Psc	MS	01h 14m	+07° 35'	5.2	23"
55 Psc		Psc	MS	00h 40m	+21° 26'	5.4	6"
65 Psc		Psc	MS	00h 50m	+27° 43'	7.0	4"
TX Psc		Psc	Var/CS	23h 46m	+03° 29'	4.5-5.3	N/A
M 45	Pleiades	Tau	OC	03h 47m	+24° 07'	1.5	120'
Melotte 25	Hyades	Tau	OC	04h 27m	+15° 52'	0.8	330'
NGC 1647	Pirate Moon Cluster	Tau	OC	04h 46m	+19° 07'	6.2	40'
M 33	Triangulum Galaxy	Tri	Gx	01h 34m	+30° 40'	6.4	62'
Alp UMi	Polaris	UMi	MS	02h 51m	+89° 20'	2.0	18"

Chart 3

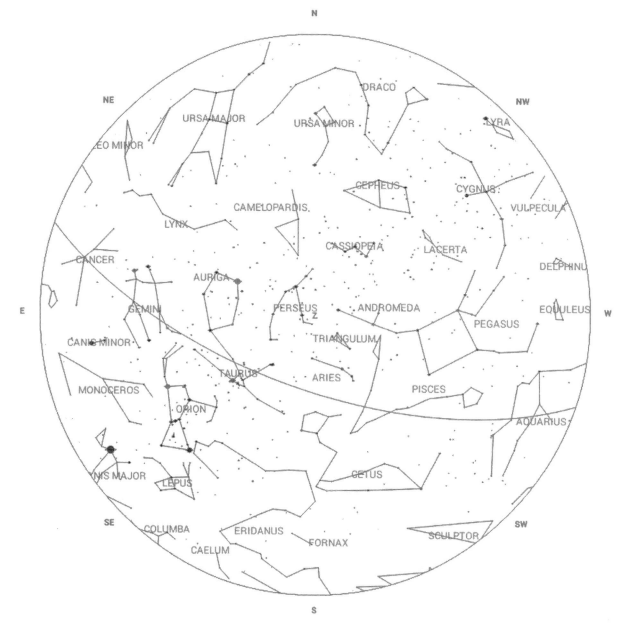

Designation	Name	Con.	Type	R.A.	Dec.	Mag	Size/Sep
Gam And	Almach	And	MS	02h 04m	+42° 20'	2.1	10"
M 31	Andromeda Galaxy	And	Gx	00h 43m	+41° 16'	4.3	156'
Pi And		And	MS	00h 38m	+33° 49'	4.4	36"
Lam Ari		Ari	MS	01h 59m	+23° 41'	4.8	37"
Gam Ari	Mesarthim	Ari	MS	01h 54m	+19° 22'	4.6	8"
30 Ari		Ari	MS	02h 37m	+24° 39'	6.5	39"
M 37		Aur	OC	05h 52m	+32° 33'	6.2	14'
M 36		Aur	OC	05h 36m	+34° 08'	6.5	10'

Designation	Name	Con.	Type	R.A.	Dec.	Mag	Size/Sep
14 Aur		Aur	MS	05h 15m	+32° 41'	5.0	15"
Kemble 1	Kemble's Cascade	Cam	Ast	03h 57m	+63° 04'	5.0	180'
NGC 1502	Jolly Roger Cluster	Cam	OC	04h 08m	+62° 20'	4.1	8'
NGC 457	Owl Cluster	Cas	OC	01h 20m	+58° 17'	5.1	20'
Iot Cas		Cas	MS	02h 29m	+67° 24'	4.5	7"
Struve 3053		Cas	MS	00h 03m	+66° 06'	5.9	15"
Eta Cas	Achird	Cas	MS	00h 50m	+57° 54'	3.6	13"
NGC 663		Cas	OC	01h 46m	+61° 14'	6.4	14'
Gam Cet	Kaffajidhma	Cet	MS	02h 43m	+03° 14'	3.5	3"
Omi Cet	Mira	Cet	Var	02h 19m	-02° 59'	2.0-10.1	N/A
32 Eri		Eri	MS	03h 54m	-02° 57'	4.4	7"
40 Eri	Keid	Eri	MS	04h 15m	-07° 39'	4.4	83"
Sig Ori		Ori	MS	05h 40m	-02° 36'	3.8	42"
Bet Ori	Rigel	Ori	MS	05h 14m	-08° 12'	0.2	10"
Alp Ori	Betelgeuse	Ori	Var	05h 55m	+07° 24'	0.4-1.3	N/A
Eta Ori		Ori	MS	05h 25m	-02° 24'	3.3	2"
Zet Ori	Alnitak	Ori	MS	05h 41m	-01° 57'	1.8	3"
23 Ori		Ori	MS	05h 23m	+03° 33'	5.0	32"
Iot Ori	Nair al Saif	Ori	MS	05h 35m	-05° 55'	2.8	11"
Collinder 70	Epsilon Orionis Cluster	Ori	OC	05h 36m	-01° 00'	0.4	150'
Struve 747		Ori	MS	05h 35m	-05° 55'	4.8	36"
Collinder 72		Ori	OC	05h 35m	-05° 55'	2.5	20'
Del Ori	Mintaka	Ori	MS	05h 33m	-00° 18'	2.1	53"
Lam Ori	Meissa	Ori	MC	05h 36m	+09° 56'	3.4	4"
M 42	Orion Nebula	Ori	Neb	05h 35m	-05° 23'	4.0	40'
NGC 1981	Coal Car Cluster	Ori	OC	05h 35m	-04° 26'	4.2	28'
Collinder 69	Lambda Orionis Cluster	Ori	OC	05h 35m	+09° 56'	2.8	70'
NGC 1528	m & m Double Cluster	Per	OC	04h 15m	+51° 13'	6.4	16'
Bet Per	Algol	Per	Var	03h 08m	+40° 57'	2.1-3.4	N/A
Eps Per		Per	MS	03h 58m	+40° 01'	2.9	9"
M 34		Per	OC	02h 42m	+42° 46'	5.8	35'
NGC 869/884	Double Cluster	Per	OC	02h 21m	+57° 08'	4.4	18'
Eta Per		Per	MS	02h 51m	+55° 54'	3.8	29"
Melotte 20	Alpha Persei Moving Cluster	Per	OC	03h 24m	+49° 52'	2.3	300'
Alp Psc	Alrisha	Psc	MS	02h 02m	+02° 46'	3.8	2"
Psi1 Psc		Psc	MS	01h 06m	+21° 28'	5.3	30"
Zet Psc		Psc	MS	01h 14m	+07° 35'	5.2	23"
55 Psc		Psc	MS	00h 40m	+21° 26'	5.4	6"
M 45	Pleiades	Tau	OC	03h 47m	+24° 07'	1.5	120'
Melotte 25	Hyades	Tau	OC	04h 27m	+15° 52'	0.8	330'
NGC 1647	Pirate Moon Cluster	Tau	OC	04h 46m	+19° 07'	6.2	40'
118 Tau		Tau	MS	05h 29m	+25° 09'	5.5	5"
M 33	Triangulum Galaxy	Tri	Gx	01h 34m	+30° 40'	6.4	62'
Alp UMi	Polaris	UMi	MS	02h 51m	+89° 20'	2.0	18"

Chart 4

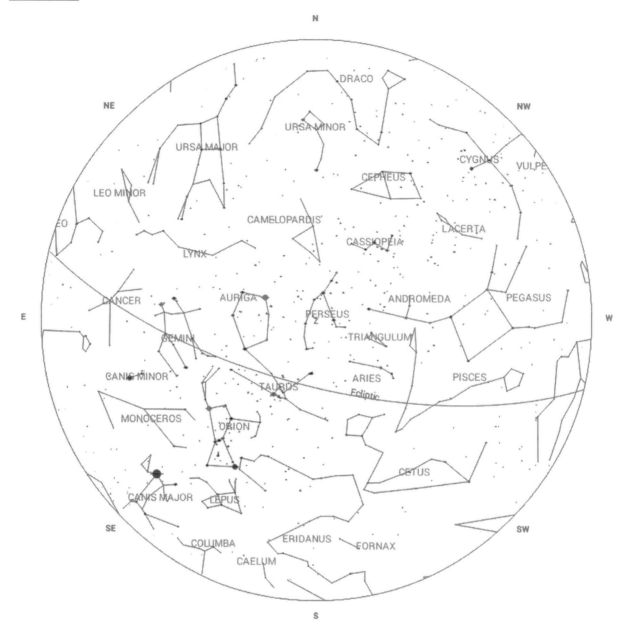

Designation	Name	Con.	Type	R.A.	Dec.	Mag	Size/Sep
Gam And	Almach	And	MS	02h 04m	+42° 20'	2.1	10"
Lam Ari		Ari	MS	01h 59m	+23° 41'	4.8	37"
Gam Ari	Mesarthim	Ari	MS	01h 54m	+19° 22'	4.6	8"
UU Aur		Aur	Var/CS	06h 36m	+38° 27'	5.3-6.5	N/A
The Aur		Aur	MS	06h 00m	+37° 13'	2.7	4"
M 37		Aur	OC	05h 52m	+32° 33'	6.2	14'
14 Aur		Aur	MS	05h 15m	+32° 41'	5.0	15"
Kemble 1	Kemble's Cascade	Cam	Ast	03h 57m	+63° 04'	5.0	180'

Designation	Name	Con.	Type	R.A.	Dec.	Mag	Size/Sep
NGC 1502	Jolly Roger Cluster	Cam	OC	04h 08m	+62° 20'	4.1	8'
NGC 457	Owl Cluster	Cas	OC	01h 20m	+58° 17'	5.1	20'
Iot Cas		Cas	MS	02h 29m	+67° 24'	4.5	7"
NGC 663		Cas	OC	01h 46m	+61° 14'	6.4	14'
Gam Cet	Kaffajidhma	Cet	MS	02h 43m	+03° 14'	3.5	3"
Omi Cet	Mira	Cet	Var	02h 19m	-02° 59'	2.0-10.1	N/A
32 Eri		Eri	MS	03h 54m	-02° 57'	4.4	7"
40 Eri	Keid	Eri	MS	04h 15m	-07° 39'	4.4	83"
M 35		Gem	OC	06h 09m	+24° 21'	5.6	25'
38 Gem		Gem	MS	06h 55m	+13° 11'	4.7	7"
Gam Lep		Lep	MS	05h 45m	-22° 27'	3.6	98"
R Lep	Hind's Crimson Star	Lep	Var/CS	05h 00m	-14° 48'	5.5-11.7	N/A
12 Lyn		Lyn	MS	06h 46m	+59° 27'	4.9	9"
Sig Ori		Ori	MS	05h 40m	-02° 36'	3.8	42"
Bet Ori	Rigel	Ori	MS	05h 14m	-08° 12'	0.2	10"
Alp Ori	Betelgeuse	Ori	Var	05h 55m	+07° 24'	0.4-1.3	N/A
Eta Ori		Ori	MS	05h 25m	-02° 24'	3.3	2"
Zet Ori	Alnitak	Ori	MS	05h 41m	-01° 57'	1.8	3"
23 Ori		Ori	MS	05h 23m	+03° 33'	5.0	32"
Iot Ori	Nair al Saif	Ori	MS	05h 35m	-05° 55'	2.8	11"
Collinder 70	Epsilon Orionis Cluster	Ori	OC	05h 36m	-01° 00'	0.4	150'
Struve 747		Ori	MS	05h 35m	-05° 55'	4.8	36"
Collinder 72		Ori	OC	05h 35m	-05° 55'	2.5	20'
Del Ori	Mintaka	Ori	MS	05h 33m	-00° 18'	2.1	53"
Lam Ori	Meissa	Ori	MC	05h 36m	+09° 56'	3.4	4"
M 42	Orion Nebula	Ori	Neb	05h 35m	-05° 23'	4.0	40'
NGC 1981	Coal Car Cluster	Ori	OC	05h 35m	-04° 26'	4.2	28'
Collinder 69	Lambda Orionis Cluster	Ori	OC	05h 35m	+09° 56'	2.8	70'
NGC 1528	m & m Double Cluster	Per	OC	04h 15m	+51° 13'	6.4	16'
Bet Per	Algol	Per	Var	03h 08m	+40° 57'	2.1-3.4	N/A
Eps Per		Per	MS	03h 58m	+40° 01'	2.9	9"
M 34		Per	OC	02h 42m	+42° 46'	5.8	35'
NGC 869/884	Double Cluster	Per	OC	02h 21m	+57° 08'	4.4	18'
Eta Per		Per	MS	02h 51m	+55° 54'	3.8	29"
Melotte 20	Alpha Persei Moving Cluster	Per	OC	03h 24m	+49° 52'	2.3	300'
Alp Psc	Alrisha	Psc	MS	02h 02m	+02° 46'	3.8	2"
Psi1 Psc		Psc	MS	01h 06m	+21° 28'	5.3	30"
Zet Psc		Psc	MS	01h 14m	+07° 35'	5.2	23"
M 45	Pleiades	Tau	OC	03h 47m	+24° 07'	1.5	120'
Melotte 25	Hyades	Tau	OC	04h 27m	+15° 52'	0.8	330'
NGC 1647	Pirate Moon Cluster	Tau	OC	04h 46m	+19° 07'	6.2	40'
118 Tau		Tau	MS	05h 29m	+25° 09'	5.5	5"
M 33	Triangulum Galaxy	Tri	Gx	01h 34m	+30° 40'	6.4	62'
Alp UMi	Polaris	UMi	MS	02h 51m	+89° 20'	2.0	18"

Chart 5

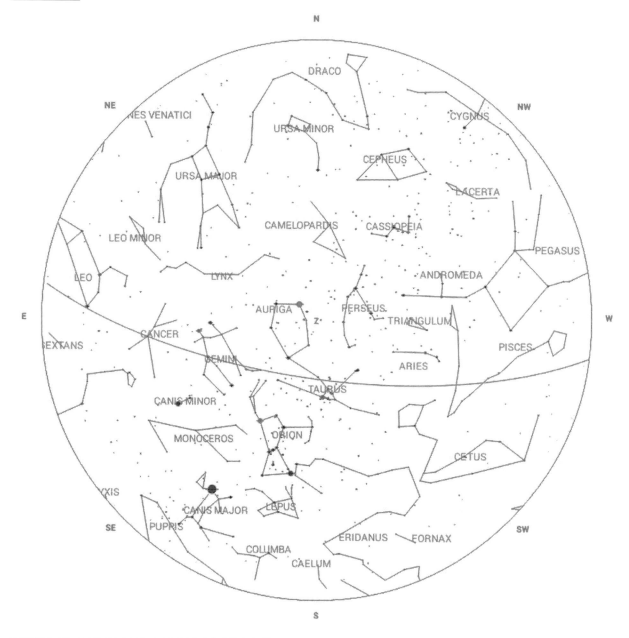

Designation	Name	Con.	Type	R.A.	Dec.	Mag	Size/Sep
Gam And	Almach	And	MS	02h 04m	+42° 20'	2.1	10"
The Aur		Aur	MS	06h 00m	+37° 13'	2.7	4"
14 Aur		Aur	MS	05h 15m	+32° 41'	5.0	15"
U Cam		Cam	Var/CS	03h 42m	+62° 39'	7.0-7.5	N/A
ST Cam		Cam	Var/CS	04h 51m	+68° 10'	7.0-8.4	N/A
Kemble 1	Kemble's Cascade	Cam	Ast	03h 57m	+63° 04'	5.0	180'
NGC 1502	Jolly Roger Cluster	Cam	OC	04h 08m	+62° 20'	4.1	8'
Iot Cas		Cas	MS	02h 29m	+67° 24'	4.5	7"

Designation	Name	Con.	Type	R.A.	Dec.	Mag	Size/Sep
32 Eri		Eri	MS	03h 54m	-02° 57'	4.4	7"
40 Eri	Keid	Eri	MS	04h 15m	-07° 39'	4.4	83"
M 35		Gem	OC	06h 09m	+24° 21'	5.6	25'
38 Gem		Gem	MS	06h 55m	+13° 11'	4.7	7"
Alp Gem	Castor	Gem	MS	07h 35m	+31° 53'	1.6	3"
Del Gem	Wasat	Gem	MS	07h 20m	+21° 59'	3.5	6"
Kap Gem		Gem	MS	07h 44m	+24° 24'	3.6	7"
Gam Lep		Lep	MS	05h 45m	-22° 27'	3.6	98"
R Lep	Hind's Crimson Star	Lep	Var/CS	05h 00m	-14° 48'	5.5-11.7	N/A
12 Lyn		Lyn	MS	06h 46m	+59° 27'	4.9	9"
19 Lyn		Lyn	MS	07h 23m	+55° 17'	5.8	215"
NGC 2353		Mon	OC	07h 15m	-10° 16'	5.2	18'
Bet Mon		Mon	MS	06h 29m	-07° 02'	3.8	7"
NGC 2237	Rosette Nebula	Mon	Neb	06h 32m	+04° 59'	5.5	70'
NGC 2244		Mon	OC	06h 32m	+04° 57'	5.2	29'
NGC 2264	Christmas Tree Cluster	Mon	OC	06h 41m	+09° 54'	4.1	39'
Eps Mon		Mon	MS	06h 24m	+04° 36'	4.3	13"
Sig Ori		Ori	MS	05h 40m	-02° 36'	3.8	42"
Bet Ori	Rigel	Ori	MS	05h 14m	-08° 12'	0.2	10"
Alp Ori	Betelgeuse	Ori	Var	05h 55m	+07° 24'	0.4-1.3	N/A
Eta Ori		Ori	MS	05h 25m	-02° 24'	3.3	2"
Zet Ori	Alnitak	Ori	MS	05h 41m	-01° 57'	1.8	3"
23 Ori		Ori	MS	05h 23m	+03° 33'	5.0	32"
Iot Ori	Nair al Saif	Ori	MS	05h 35m	-05° 55'	2.8	11"
W Ori		Ori	Var/CS	05h 05m	+01° 11'	6.2-7.0	N/A
Collinder 70	Epsilon Orionis Cluster	Ori	OC	05h 36m	-01° 00'	0.4	150'
Struve 747		Ori	MS	05h 35m	-05° 55'	4.8	36"
Collinder 72		Ori	OC	05h 35m	-05° 55'	2.5	20'
Del Ori	Mintaka	Ori	MS	05h 33m	-00° 18'	2.1	53"
Lam Ori	Meissa	Ori	MC	05h 36m	+09° 56'	3.4	4"
M 42	Orion Nebula	Ori	Neb	05h 35m	-05° 23'	4.0	40'
NGC 1981	Coal Car Cluster	Ori	OC	05h 35m	-04° 26'	4.2	28'
Collinder 69	Lambda Orionis Cluster	Ori	OC	05h 35m	+09° 56'	2.8	70'
BL Ori		Ori	Var/CS	06h 26m	+14° 43'	6.3-7.0	N/A
Bet Per	Algol	Per	Var	03h 08m	+40° 57'	2.1-3.4	N/A
Eps Per		Per	MS	03h 58m	+40° 01'	2.9	9"
M 34		Per	OC	02h 42m	+42° 46'	5.8	35'
NGC 869/884	Double Cluster	Per	OC	02h 21m	+57° 08'	4.4	18'
Eta Per		Per	MS	02h 51m	+55° 54'	3.8	29"
Melotte 20	Alpha Persei Moving Cluster	Per	OC	03h 24m	+49° 52'	2.3	300'
M 45	Pleiades	Tau	OC	03h 47m	+24° 07'	1.5	120'
Melotte 25	Hyades	Tau	OC	04h 27m	+15° 52'	0.8	330'
118 Tau		Tau	MS	05h 29m	+25° 09'	5.5	5"
Alp UMi	Polaris	UMi	MS	02h 51m	+89° 20'	2.0	18"

Chart 6

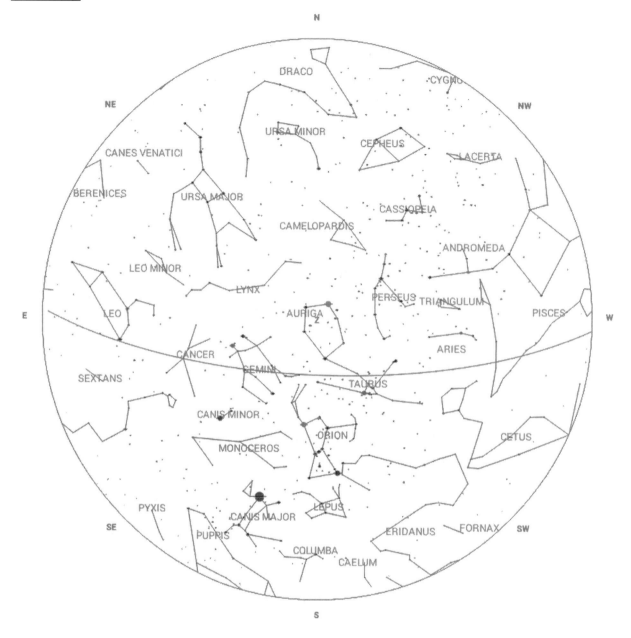

Designation	Name	Con.	Type	R.A.	Dec.	Mag	Size/Sep
UU Aur		Aur	Var/CS	06h 36m	+38° 27'	5.3-6.5	N/A
The Aur		Aur	MS	06h 00m	+37° 13'	2.7	4"
14 Aur		Aur	MS	05h 15m	+32° 41'	5.0	15"
Kemble 1	Kemble's Cascade	Cam	Ast	03h 57m	+63° 04'	5.0	180'
NGC 1502	Jolly Roger Cluster	Cam	OC	04h 08m	+62° 20'	4.1	8'
M 41		CMa	OC	05h 46m	-20° 45'	5.0	39'
Eps CMa	Adhara	CMa	MS	06h 59m	-28° 58'	1.5	8"
NGC 2362	Tau Canis Majoris Cluster	CMa	OC	07h 19m	-24° 57'	3.8	5'

Designation	Name	Con.	Type	R.A.	Dec.	Mag	Size/Sep
Iot Cnc		Cnc	MS	08h 47m	+28° 46'	4.0	30"
M 44	Praesepe	Cnc	OC	08h 40m	+19° 40'	3.9	70'
Zet Cnc	Tegmen	Cnc	MS	08h 12m	+17° 39'	4.7	6"
Phi2 Cnc		Cnc	MS	08h 27m	+26° 56'	5.6	5"
57 Cnc		Cnc	MS	08h 54m	+30° 35'	5.4	56"
X Cnc		Cnc	Var/CS	08h 55m	+17° 14'	5.6-7.5	N/A
32 Eri		Eri	MS	03h 54m	-02° 57'	4.4	7"
40 Eri	Keid	Eri	MS	04h 15m	-07° 39'	4.4	83"
M 35		Gem	OC	06h 09m	+24° 21'	5.6	25'
38 Gem		Gem	MS	06h 55m	+13° 11'	4.7	7"
Alp Gem	Castor	Gem	MS	07h 35m	+31° 53'	1.6	3"
Del Gem	Wasat	Gem	MS	07h 20m	+21° 59'	3.5	6"
Kap Gem		Gem	MS	07h 44m	+24° 24'	3.6	7"
Gam Lep		Lep	MS	05h 45m	-22° 27'	3.6	98"
R Lep	Hind's Crimson Star	Lep	Var/CS	05h 00m	-14° 48'	5.5-11.7	N/A
12 Lyn		Lyn	MS	06h 46m	+59° 27'	4.9	9"
19 Lyn		Lyn	MS	07h 23m	+55° 17'	5.8	215"
NGC 2353		Mon	OC	07h 15m	-10° 16'	5.2	18'
Bet Mon		Mon	MS	06h 29m	-07° 02'	3.8	7"
NGC 2237	Rosette Nebula	Mon	Neb	06h 32m	+04° 59'	5.5	70'
NGC 2244		Mon	OC	06h 32m	+04° 57'	5.2	29'
NGC 2264	Christmas Tree Cluster	Mon	OC	06h 41m	+09° 54'	4.1	39'
Eps Mon		Mon	MS	06h 24m	+04° 36'	4.3	13"
Sig Ori		Ori	MS	05h 40m	-02° 36'	3.8	42"
Bet Ori	Rigel	Ori	MS	05h 14m	-08° 12'	0.2	10"
Alp Ori	Betelgeuse	Ori	Var	05h 55m	+07° 24'	0.4-1.3	N/A
Eta Ori		Ori	MS	05h 25m	-02° 24'	3.3	2"
Zet Ori	Alnitak	Ori	MS	05h 41m	-01° 57'	1.8	3"
23 Ori		Ori	MS	05h 23m	+03° 33'	5.0	32"
Iot Ori	Nair al Saif	Ori	MS	05h 35m	-05° 55'	2.8	11"
Collinder 70	Epsilon Orionis Cluster	Ori	OC	05h 36m	-01° 00'	0.4	150'
Struve 747		Ori	MS	05h 35m	-05° 55'	4.8	36"
Collinder 72		Ori	OC	05h 35m	-05° 55'	2.5	20'
Del Ori	Mintaka	Ori	MS	05h 33m	-00° 18'	2.1	53"
Lam Ori	Meissa	Ori	MC	05h 36m	+09° 56'	3.4	4"
M 42	Orion Nebula	Ori	Neb	05h 35m	-05° 23'	4.0	40'
NGC 1981	Coal Car Cluster	Ori	OC	05h 35m	-04° 26'	4.2	28'
Collinder 69	Lambda Orionis Cluster	Ori	OC	05h 35m	+09° 56'	2.8	70'
Bet Per	Algol	Per	Var	03h 08m	+40° 57'	2.1-3.4	N/A
Eps Per		Per	MS	03h 58m	+40° 01'	2.9	9"
Melotte 20	Alpha Persei Moving Cluster	Per	OC	03h 24m	+49° 52'	2.3	300'
M 45	Pleiades	Tau	OC	03h 47m	+24° 07'	1.5	120'
Melotte 25	Hyades	Tau	OC	04h 27m	+15° 52'	0.8	330'
118 Tau		Tau	MS	05h 29m	+25° 09'	5.5	5"

Chart 7

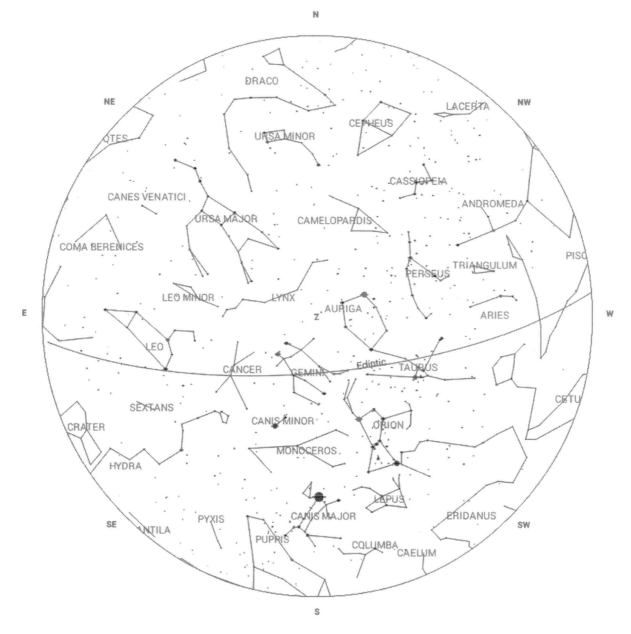

Designation	Name	Con.	Type	R.A.	Dec.	Mag	Size/Sep
14 Aur		Aur	MS	05h 15m	+32° 41'	5.0	15"
The Aur		Aur	MS	06h 00m	+37° 13'	2.7	4"
M 37		Aur	OC	05h 52m	+32° 33'	6.2	14'
UU Aur		Aur	Var/CS	06h 36m	+38° 27'	5.3-6.5	N/A
NGC 1502	Jolly Roger Cluster	Cam	OC	04h 08m	+62° 20'	4.1	8'
Eps CMa	Adhara	CMa	MS	06h 59m	-28° 58'	1.5	8"
M 41		CMa	OC	05h 46m	-20° 45'	5.0	39'
NGC 2362	Tau Canis Majoris Cluster	CMa	OC	07h 19m	-24° 57'	3.8	5'

Designation	Name	Con.	Type	R.A.	Dec.	Mag	Size/Sep
Zet Cnc	Tegmen	Cnc	MS	08h 12m	+17° 39'	4.7	6"
Phi2 Cnc		Cnc	MS	08h 27m	+26° 56'	5.6	5"
Iot Cnc		Cnc	MS	08h 47m	+28° 46'	4.0	30"
57 Cnc		Cnc	MS	08h 54m	+30° 35'	5.4	56"
M 44	Praesepe	Cnc	OC	08h 40m	+19° 40'	3.9	70'
X Cnc		Cnc	Var/CS	08h 55m	+17° 14'	5.6-7.5	N/A
38 Gem		Gem	MS	06h 55m	+13° 11'	4.7	7"
Del Gem	Wasat	Gem	MS	07h 20m	+21° 59'	3.5	6"
Alp Gem	Castor	Gem	MS	07h 35m	+31° 53'	1.6	3"
Kap Gem		Gem	MS	07h 44m	+24° 24'	3.6	7"
M 35		Gem	OC	06h 09m	+24° 21'	5.6	25'
R Leo	Peltier's Variable Star	Leo	Var	09h 48m	+11° 26'	4.4-10.5	N/A
Gam Lep		Lep	MS	05h 45m	-22° 27'	3.6	98"
R Lep	Hind's Crimson Star	Lep	Var/CS	05h 00m	-14° 48'	5.5-11.7	N/A
12 Lyn		Lyn	MS	06h 46m	+59° 27'	4.9	9"
19 Lyn		Lyn	MS	07h 23m	+55° 17'	5.8	215"
38 Lyn		Lyn	MS	09h 19m	+36° 48'	3.8	3"
Eps Mon		Mon	MS	06h 24m	+04° 36'	4.3	13"
Bet Mon		Mon	MS	06h 29m	-07° 02'	3.8	7"
NGC 2237	Rosette Nebula	Mon	Neb	06h 32m	+04° 59'	5.5	70'
NGC 2244		Mon	OC	06h 32m	+04° 57'	5.2	29'
NGC 2264	Christmas Tree Cluster	Mon	OC	06h 41m	+09° 54'	4.1	39'
NGC 2301	Hagrid's Dragon	Mon	OC	06h 52m	+00° 28'	6.3	14'
NGC 2353		Mon	OC	07h 15m	-10° 16'	5.2	18'
Lam Ori	Meissa	Ori	MC	05h 36m	+09° 56'	3.4	4"
Bet Ori	Rigel	Ori	MS	05h 14m	-08° 12'	0.2	10"
23 Ori		Ori	MS	05h 23m	+03° 33'	5.0	32"
Eta Ori		Ori	MS	05h 25m	-02° 24'	3.3	2"
Del Ori	Mintaka	Ori	MS	05h 33m	-00° 18'	2.1	53"
Iot Ori	Nair al Saif	Ori	MS	05h 35m	-05° 55'	2.8	11"
Struve 747		Ori	MS	05h 35m	-05° 55'	4.8	36"
Sig Ori		Ori	MS	05h 40m	-02° 36'	3.8	42"
Zet Ori	Alnitak	Ori	MS	05h 41m	-01° 57'	1.8	3"
M 42	Orion Nebula	Ori	Neb	05h 35m	-05° 23'	4.0	40'
Collinder 72		Ori	OC	05h 35m	-05° 55'	2.5	20'
NGC 1981	Coal Car Cluster	Ori	OC	05h 35m	-04° 26'	4.2	28'
Collinder 69	Lambda Orionis Cluster	Ori	OC	05h 35m	+09° 56'	2.8	70'
Collinder 70	Epsilon Orionis Cluster	Ori	OC	05h 36m	-01° 00'	0.4	150'
Alp Ori	Betelgeuse	Ori	Var	05h 55m	+07° 24'	0.4-1.3	N/A
M 47		Pup	OC	07h 37m	-14° 29'	4.3	25'
118 Tau		Tau	MS	05h 29m	+25° 09'	5.5	5"
Melotte 25	Hyades	Tau	OC	04h 27m	+15° 52'	0.8	330'
NGC 1647	Pirate Moon Cluster	Tau	OC	04h 46m	+19° 07'	6.2	40'

Chart 8

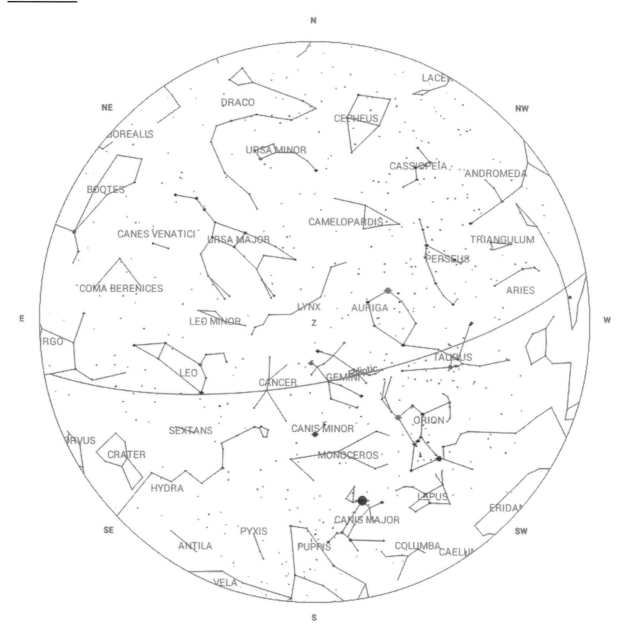

Designation	Name	Con.	Type	R.A.	Dec.	Mag	Size/Sep
14 Aur		Aur	MS	05h 15m	+32° 41'	5.0	15"
The Aur		Aur	MS	06h 00m	+37° 13'	2.7	4"
M 37		Aur	OC	05h 52m	+32° 33'	6.2	14'
UU Aur		Aur	Var/CS	06h 36m	+38° 27'	5.3-6.5	N/A
Eps CMa	Adhara	CMa	MS	06h 59m	-28° 58'	1.5	8"
M 41		CMa	OC	05h 46m	-20° 45'	5.0	39'
NGC 2362	Tau Canis Majoris Cluster	CMa	OC	07h 19m	-24° 57'	3.8	5'
Zet Cnc	Tegmen	Cnc	MS	08h 12m	+17° 39'	4.7	6"

Designation	Name	Con.	Type	R.A.	Dec.	Mag	Size/Sep
Phi2 Cnc		Cnc	MS	08h 27m	+26° 56'	5.6	5"
Iot Cnc		Cnc	MS	08h 47m	+28° 46'	4.0	30"
57 Cnc		Cnc	MS	08h 54m	+30° 35'	5.4	56"
M 44	Praesepe	Cnc	OC	08h 40m	+19° 40'	3.9	70'
X Cnc		Cnc	Var/CS	08h 55m	+17° 14'	5.6-7.5	N/A
38 Gem		Gem	MS	06h 55m	+13° 11'	4.7	7"
Del Gem	Wasat	Gem	MS	07h 20m	+21° 59'	3.5	6"
Alp Gem	Castor	Gem	MS	07h 35m	+31° 53'	1.6	3"
Kap Gem		Gem	MS	07h 44m	+24° 24'	3.6	7"
M 35		Gem	OC	06h 09m	+24° 21'	5.6	25'
Eps Hya		Hya	MS	08h 47m	+06° 25'	3.4	3"
M 48		Hya	OC	08h 14m	-05° 45'	5.5	30'
U Hya		Hya	Var/CS	10h 38m	-13° 23'	4.8-6.5	N/A
R Leo	Peltier's Variable Star	Leo	Var	09h 48m	+11° 26'	4.4-10.5	N/A
Gam Leo	Algieba	Leo	MS	10h 20m	+19° 50'	2.0	5"
54 Leo		Leo	MS	10h 56m	+24° 45'	4.3	6"
12 Lyn		Lyn	MS	06h 46m	+59° 27'	4.9	9"
19 Lyn		Lyn	MS	07h 23m	+55° 17'	5.8	215"
38 Lyn		Lyn	MS	09h 19m	+36° 48'	3.8	3"
Eps Mon		Mon	MS	06h 24m	+04° 36'	4.3	13"
Bet Mon		Mon	MS	06h 29m	-07° 02'	3.8	7"
NGC 2237	Rosette Nebula	Mon	Neb	06h 32m	+04° 59'	5.5	70'
NGC 2244		Mon	OC	06h 32m	+04° 57'	5.2	29'
NGC 2264	Christmas Tree Cluster	Mon	OC	06h 41m	+09° 54'	4.1	39'
NGC 2301	Hagrid's Dragon	Mon	OC	06h 52m	+00° 28'	6.3	14'
NGC 2353		Mon	OC	07h 15m	-10° 16'	5.2	18'
Lam Ori	Meissa	Ori	MC	05h 36m	+09° 56'	3.4	4"
Bet Ori	Rigel	Ori	MS	05h 14m	-08° 12'	0.2	10"
23 Ori		Ori	MS	05h 23m	+03° 33'	5	32"
Eta Ori		Ori	MS	05h 25m	-02° 24'	3.3	2"
Del Ori	Mintaka	Ori	MS	05h 33m	-00° 18'	2.1	53"
Iot Ori	Nair al Saif	Ori	MS	05h 35m	-05° 55'	2.8	11"
Struve 747		Ori	MS	05h 35m	-05° 55'	4.8	36"
Sig Ori		Ori	MS	05h 40m	-02° 36'	3.8	42"
Zet Ori	Alnitak	Ori	MS	05h 41m	-01° 57'	1.8	3"
M 42	Orion Nebula	Ori	Neb	05h 35m	-05° 23'	4.0	40'
Collinder 72		Ori	OC	05h 35m	-05° 55'	2.5	20'
NGC 1981	Coal Car Cluster	Ori	OC	05h 35m	-04° 26'	4.2	28'
Collinder 69	Lambda Orionis Cluster	Ori	OC	05h 35m	+09° 56'	2.8	70'
Collinder 70	Epsilon Orionis Cluster	Ori	OC	05h 36m	-01° 00'	0.4	150'
Alp Ori	Betelgeuse	Ori	Var	05h 55m	+07° 24'	0.4-1.3	N/A
M 47		Pup	OC	07h 37m	-14° 29'	4.3	25'
118 Tau		Tau	MS	05h 29m	+25° 09'	5.5	5"

Chart 9

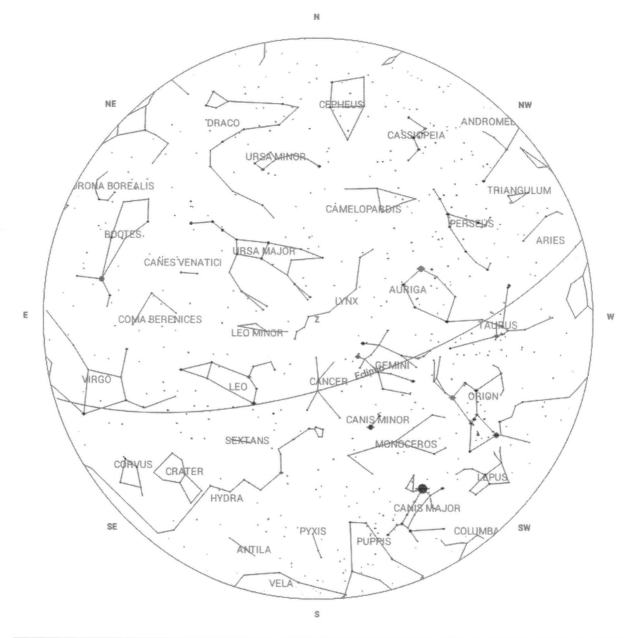

Designation	Name	Con.	Type	R.A.	Dec.	Mag	Size/Sep
The Aur		Aur	MS	06h 00m	+37° 13'	2.7	4"
UU Aur		Aur	Var/CS	06h 36m	+38° 27'	5.3-6.5	N/A
NGC 2403		Cam	Gx	07h 37m	+65° 36'	8.8	20'
Zet Cnc	Tegmen	Cnc	MS	08h 12m	+17° 39'	4.7	6"
Phi2 Cnc		Cnc	MS	08h 27m	+26° 56'	5.6	5"
Iot Cnc		Cnc	MS	08h 47m	+28° 46'	4.0	30"
57 Cnc		Cnc	MS	08h 54m	+30° 35'	5.4	56"
M 44	Praesepe	Cnc	OC	08h 40m	+19° 40'	3.9	70'

Designation	Name	Con.	Type	R.A.	Dec.	Mag	Size/Sep
M 67		Cnc	OC	08h 51m	+11° 48'	7.4	25'
X Cnc		Cnc	Var/CS	08h 55m	+17° 14'	5.6-7.5	N/A
38 Gem		Gem	MS	06h 55m	+13° 11'	4.7	7"
Del Gem	Wasat	Gem	MS	07h 20m	+21° 59'	3.5	6"
Alp Gem	Castor	Gem	MS	07h 35m	+31° 53'	1.6	3"
Kap Gem		Gem	MS	07h 44m	+24° 24'	3.6	7"
M 35		Gem	OC	06h 09m	+24° 21'	5.6	25'
NGC 2392	Eskimo Nebula	Gem	PN	07h 29m	+20° 55'	8.6	47"
NGC 3242	Ghost of Jupiter	Hya	PN	10h 25m	-18° 39'	8.6	40"
Eps Hya		Hya	MS	08h 47m	+06° 25'	3.4	3"
M 48		Hya	OC	08h 14m	-05° 45'	5.5	30'
U Hya		Hya	Var/CS	10h 38m	-13° 23'	4.8-6.5	N/A
R Leo	Peltier's Variable Star	Leo	Var	09h 48m	+11° 26'	4.4-10.5	N/A
Gam Leo	Algieba	Leo	MS	10h 20m	+19° 50'	2.0	5"
54 Leo		Leo	MS	10h 56m	+24° 45'	4.3	6"
M 66		Leo	Gx	11h 20m	+13° 00'	9.7	9'
12 Lyn		Lyn	MS	06h 46m	+59° 27'	4.9	9"
19 Lyn		Lyn	MS	07h 23m	+55° 17'	5.8	215"
38 Lyn		Lyn	MS	09h 19m	+36° 48'	3.8	3"
Eps Mon		Mon	MS	06h 24m	+04° 36'	4.3	13"
NGC 3521		Leo	Gx	11h 06m	-00° 02'	9.9	10'
Bet Mon		Mon	MS	06h 29m	-07° 02'	3.8	7"
Iot Leo		Leo	MS	11h 24m	+10° 32'	3.9	2"
NGC 2237	Rosette Nebula	Mon	Neb	06h 32m	+04° 59'	5.5	70'
NGC 2244		Mon	OC	06h 32m	+04° 57'	5.2	29'
NGC 2264	Christmas Tree Cluster	Mon	OC	06h 41m	+09° 54'	4.1	39'
NGC 2301	Hagrid's Dragon	Mon	OC	06h 52m	+00° 28'	6.3	14'
M 50		Mon	OC	07h 03m	-08° 23'	7.2	14'
NGC 2353		Mon	OC	07h 15m	-10° 16'	5.2	18'
NGC 2506		Mon	OC	08h 00m	-10° 46'	8.9	12'
NGC 2175		Ori	Neb	06h 10m	+20° 29'	6.8	22'
NGC 2169	37 Cluster	Ori	OC	06h 09m	+13° 58'	7.0	5'
NGC 2467		Pup	Neb	07h 52m	-26° 26'	7.1	14'
M 47		Pup	OC	07h 37m	-14° 29'	4.3	25'
M 46		Pup	OC	07h 42m	-14° 48'	6.6	20'
M 93		Pup	OC	07h 45m	-23° 51'	6.5	10'
NGC 2539	Dish Cluster	Pup	OC	08h 11m	-12° 49'	8.0	9'
M 97	Owl Nebula	UMa	PN	11h 15m	+55° 01'	9.7	3'
Xi UMa	Alula Australis	UMa	MS	11h 18m	+31° 32'	4.4	2"
VY UMa		UMa	Var/CS	10h 45m	+67° 25'	5.9-6.5	N/A
M 81	Bode's Galaxy	UMa	Gx	09h 56m	+69° 04'	7.8	22'
M 82	Cigar Galaxy	UMa	Gx	09h 56m	+69° 41'	9.0	9'

Chart 10

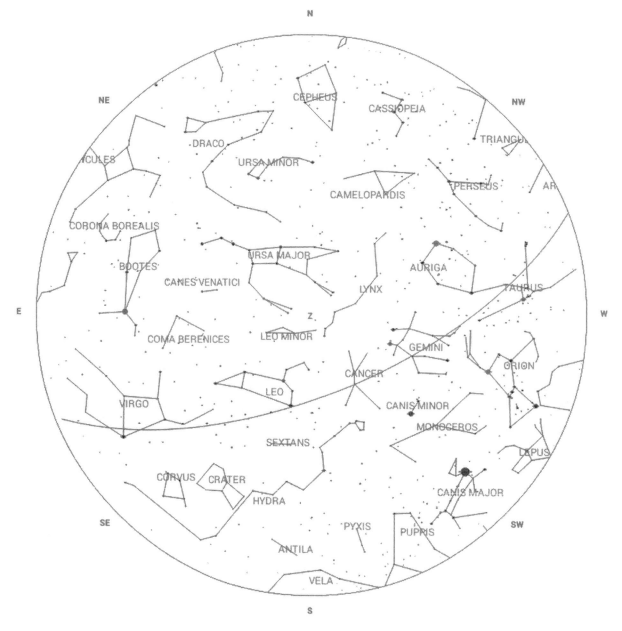

Designation	Name	Con.	Type	R.A.	Dec.	Mag	Size/Sep
NGC 2403		Cam	Gx	07h 37m	+65° 36'	8.8	20'
Zet Cnc	Tegmen	Cnc	MS	08h 12m	+17° 39'	4.7	6"
Phi2 Cnc		Cnc	MS	08h 27m	+26° 56'	5.6	5"
Iot Cnc		Cnc	MS	08h 47m	+28° 46'	4.0	30"
57 Cnc		Cnc	MS	08h 54m	+30° 35'	5.4	56"
M 44	Praesepe	Cnc	OC	08h 40m	+19° 40'	3.9	70'
M 67		Cnc	OC	08h 51m	+11° 48'	7.4	25'
X Cnc		Cnc	Var/CS	08h 55m	+17° 14'	5.6-7.5	N/A
M 64	Black Eye Galaxy	Com	Gx	12h 57m	+21° 41'	9.3	10'

Designation	Name	Con.	Type	R.A.	Dec.	Mag	Size/Sep
Melotte 111	Coma Star Cluster	Com	OC	12h 25m	+26° 06'	2.9	120'
NGC 4725		Com	Gx	12h 50m	+25° 30'	9.9	10'
24 Com		Com	MS	12h 35m	+18° 23'	5.0	20"
M 106		CVn	Gx	12h 19m	+47° 18'	9.1	17'
Alp CVn	Cor Caroli	CVn	MS	12h 56m	+38° 19'	2.9	19"
M 94		CVn	Gx	12h 51m	+41° 07'	8.7	10'
2 CVn		CVn	MS	12h 16m	+40° 40'	5.7	11"
NGC 4656	Hook Galaxy	CVn	Gx	12h 44m	+32° 10'	9.7	9'
NGC 4449		CVn	Gx	12h 28m	+44° 06'	9.5	5'
Y CVn	La Superba	CVn	CS	12h 45m	+45° 26'	5.2-5.5	N/A
NGC 4490	Cocoon Galaxy	CVn	Gx	12h 31m	+41° 39'	9.8	6'
NGC 4631	Whale Galaxy	CVn	Gx	12h 42m	+32° 33'	9.5	13'
RY Dra		Dra	CS	12h 56m	+66° 00'	6.0-8.0	N/A
Del Gem	Wasat	Gem	MS	07h 20m	+21° 59'	3.5	6"
Alp Gem	Castor	Gem	MS	07h 35m	+31° 53'	1.6	3"
Kap Gem		Gem	MS	07h 44m	+24° 24'	3.6	7"
NGC 2392	Eskimo Nebula	Gem	PN	07h 29m	+20° 55'	8.6	47"
M 68		Hya	GC	12h 39m	-26° 45'	7.3	11'
NGC 3242	Ghost of Jupiter	Hya	PN	10h 25m	-18° 39'	8.6	40"
Eps Hya		Hya	MS	08h 47m	+06° 25'	3.4	3"
M 48		Hya	OC	08h 14m	-05° 45'	5.5	30'
U Hya		Hya	Var/CS	10h 38m	-13° 23'	4.8-6.5	N/A
R Leo	Peltier's Variable Star	Leo	Var	09h 48m	+11° 26'	4.4-10.5	N/A
Gam Leo	Algieba	Leo	MS	10h 20m	+19° 50'	2.0	5"
54 Leo		Leo	MS	10h 56m	+24° 45'	4.3	6"
M 66		Leo	Gx	11h 20m	+13° 00'	9.7	9'
NGC 3521		Leo	Gx	11h 06m	-00° 02'	9.9	10'
Iot Leo		Leo	MS	11h 24m	+10° 32'	3.9	2"
19 Lyn		Lyn	MS	07h 23m	+55° 17'	5.8	215"
38 Lyn		Lyn	MS	09h 19m	+36° 48'	3.8	3"
M 40	Winnecke 4	UMa	MS	12h 22m	+58° 05'	9.6	
M 97	Owl Nebula	UMa	PN	11h 15m	+55° 01'	9.7	3'
Xi UMa	Alula Australis	UMa	MS	11h 18m	+31° 32'	4.4	2"
VY UMa		UMa	Var/CS	10h 45m	+67° 25'	5.9-6.5	N/A
M 81	Bode's Galaxy	UMa	Gx	09h 56m	+69° 04'	7.8	22'
M 82	Cigar Galaxy	UMa	Gx	09h 56m	+69° 41'	9.0	9'
M 87		Vir	Gx	12h 31m	+12° 23'	9.6	8'
M 104	Sombrero Galaxy	Vir	Gx	12h 40m	-11° 37'	9.1	9'
M 49		Vir	Gx	12h 30m	+08° 00'	9.3	9'
M 60		Vir	Gx	12h 44m	+11° 33'	9.8	7'
M 86		Vir	Gx	12h 26m	+12° 57'	9.8	10'
Gam Vir	Porrima	Vir	MS	12h 42m	-01° 27'	2.7	2"
SS Vir		Vir	Var/CS	12h 25m	+00° 48'	6.0-9.6	6.0-9.6

Chart 11

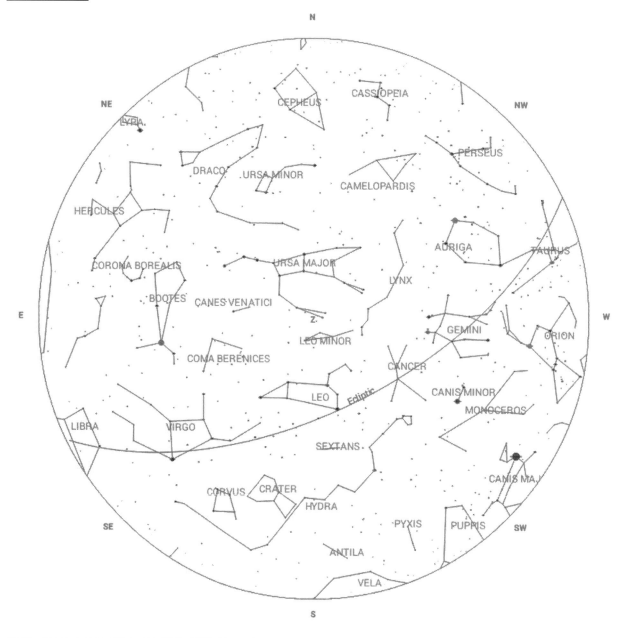

Designation	Name	Con.	Type	R.A.	Dec.	Mag	Size/Sep
Zet Cnc	Tegmen	Cnc	MS	08h 12m	+17° 39'	4.7	6"
Phi2 Cnc		Cnc	MS	08h 27m	+26° 56'	5.6	5"
Iot Cnc		Cnc	MS	08h 47m	+28° 46'	4.0	30"
57 Cnc		Cnc	MS	08h 54m	+30° 35'	5.4	56"
M 44	Praesepe	Cnc	OC	08h 40m	+19° 40'	3.9	70'
M 67		Cnc	OC	08h 51m	+11° 48'	7.4	25'
X Cnc		Cnc	Var/CS	08h 55m	+17° 14'	5.6-7.5	N/A
M 53		Com	GC	13h 13m	+18° 10'	7.7	13'
M 64	Black Eye Galaxy	Com	Gx	12h 57m	+21° 41'	9.3	10'

Designation	Name	Con.	Type	R.A.	Dec.	Mag	Size/Sep
Melotte 111	Coma Star Cluster	Com	OC	12h 25m	+26° 06'	2.9	120'
NGC 4725		Com	Gx	12h 50m	+25° 30'	9.9	10'
24 Com		Com	MS	12h 35m	+18° 23'	5.0	20"
M 51	Whirlpool Galaxy	CVn	Gx	13h 30m	+47° 12'	8.7	10'
M 63	Sunflower Galaxy	CVn	Gx	13h 16m	+42° 02'	9.3	12'
M 106		CVn	Gx	12h 19m	+47° 18'	9.1	17'
M 3		CVn	GC	13h 42m	+28° 23'	6.3	18'
Alp CVn	Cor Caroli	CVn	MS	12h 56m	+38° 19'	2.9	19"
M 94		CVn	Gx	12h 51m	+41° 07'	8.7	10'
2 CVn		CVn	MS	12h 16m	+40° 40'	5.7	11"
NGC 4656	Hook Galaxy	CVn	Gx	12h 44m	+32° 10'	9.7	9'
NGC 4449		CVn	Gx	12h 28m	+44° 06'	9.5	5'
Y CVn	La Superba	CVn	CS	12h 45m	+45° 26'	5.2-5.5	N/A
NGC 4490	Cocoon Galaxy	CVn	Gx	12h 31m	+41° 39'	9.8	6'
NGC 4631	Whale Galaxy	CVn	Gx	12h 42m	+32° 33'	9.5	13'
RY Dra		Dra	CS	12h 56m	+66° 00'	6.0-8.0	N/A
M 83		Hya	Gx	13h 37m	-29° 52'	7.8	14'
M 68		Hya	GC	12h 39m	-26° 45'	7.3	11'
NGC 3242	Ghost of Jupiter	Hya	PN	10h 25m	-18° 39'	8.6	40"
Eps Hya		Hya	MS	08h 47m	+06° 25'	3.4	3"
M 48		Hya	OC	08h 14m	-05° 45'	5.5	30'
U Hya		Hya	Var/CS	10h 38m	-13° 23'	4.8-6.5	N/A
R Leo	Peltier's Variable Star	Leo	Var	09h 48m	+11° 26'	4.4-10.5	N/A
Gam Leo	Algieba	Leo	MS	10h 20m	+19° 50'	2.0	5"
54 Leo		Leo	MS	10h 56m	+24° 45'	4.3	6"
M 66		Leo	Gx	11h 20m	+13° 00'	9.7	9'
NGC 3521		Leo	Gx	11h 06m	-00° 02'	9.9	10'
Iot Leo		Leo	MS	11h 24m	+10° 32'	3.9	2"
38 Lyn		Lyn	MS	09h 19m	+36° 48'	3.8	3"
M 40	Winnecke 4	UMa	MS	12h 22m	+58° 05'	9.6	
M 97	Owl Nebula	UMa	PN	11h 15m	+55° 01'	9.7	3'
Zet UMa	Mizar & Alcor	UMa	MS	13h 24m	+54° 56'	2.1	711"
Xi UMa	Alula Australis	UMa	MS	11h 18m	+31° 32'	4.4	2"
VY UMa		UMa	Var/CS	10h 45m	+67° 25'	5.9-6.5	N/A
M 81	Bode's Galaxy	UMa	Gx	09h 56m	+69° 04'	7.8	22'
M 82	Cigar Galaxy	UMa	Gx	09h 56m	+69° 41'	9.0	9'
M 87		Vir	Gx	12h 31m	+12° 23'	9.6	8'
M 104	Sombrero Galaxy	Vir	Gx	12h 40m	-11° 37'	9.1	9'
M 49		Vir	Gx	12h 30m	+08° 00'	9.3	9'
M 60		Vir	Gx	12h 44m	+11° 33'	9.8	7'
The Vir		Vir	MS	13h 10m	-05° 32'	4.4	70"
M 86		Vir	Gx	12h 26m	+12° 57'	9.8	10'
Gam Vir	Porrima	Vir	MS	12h 42m	-01° 27'	2.7	2"
SS Vir		Vir	Var/CS	12h 25m	+00° 48'	6.0-9.6	6.0-9.6

Chart 12

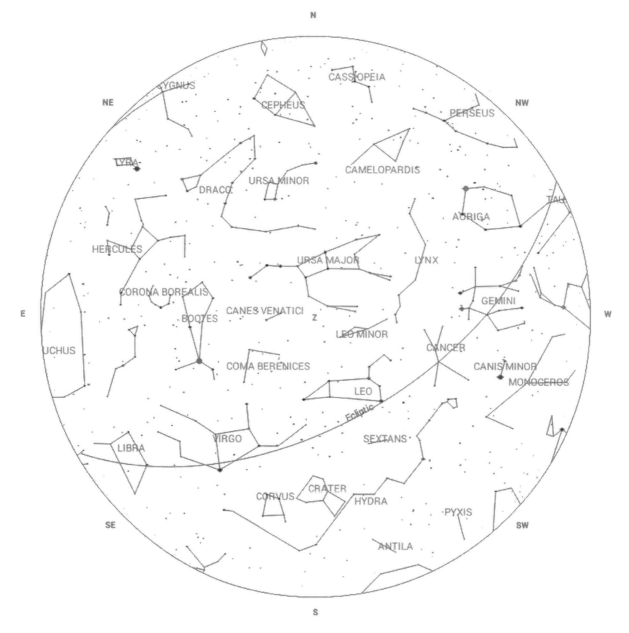

Designation	Name	Con.	Type	R.A.	Dec.	Mag	Size/Sep
Eps Boo	Izar	Boo	MS	14h 45m	+27° 04'	2.4	3"
Xi Boo		Boo	MS	14h 51m	+19° 06'	4.6	6"
Kap Boo	Asellus Tertius	Boo	MS	14h 14m	+51° 47'	4.5	13"
Pi Boo		Boo	MS	14h 41m	+16° 25'	4.5	6"
39 Boo		Boo	MS	14h 50m	+48° 43'	5.7	3"
Struve 1835		Boo	MS	14h 23m	+08° 27'	4.9	6"
M 53		Com	GC	13h 13m	+18° 10'	7.7	13'
M 64	Black Eye Galaxy	Com	Gx	12h 57m	+21° 41'	9.3	10'
Melotte 111	Coma Star Cluster	Com	OC	12h 25m	+26° 06'	2.9	120'

Designation	Name	Con.	Type	R.A.	Dec.	Mag	Size/Sep
24 Com		Com	MS	12h 35m	+18° 23'	5.0	20"
Del Crv	Algorab	Crv	MS	12h 30m	-16° 31'	5.7	24"
Struve 1669		Crv	MS	12h 41m	-13° 01'	5.2	5"
M 51	Whirlpool Galaxy	CVn	Gx	13h 30m	+47° 12'	8.7	10'
M 63	Sunflower Galaxy	CVn	Gx	13h 16m	+42° 02'	9.3	12'
M 106		CVn	Gx	12h 19m	+47° 18'	9.1	17'
M 3		CVn	GC	13h 42m	+28° 23'	6.3	18'
Alp CVn	Cor Caroli	CVn	MS	12h 56m	+38° 19'	2.9	19"
M 94		CVn	Gx	12h 51m	+41° 07'	8.7	10'
2 CVn		CVn	MS	12h 16m	+40° 40'	5.7	11"
NGC 4656	Hook Galaxy	CVn	Gx	12h 44m	+32° 10'	9.7	9'
NGC 4449		CVn	Gx	12h 28m	+44° 06'	9.5	5'
Y CVn	La Superba	CVn	CS	12h 45m	+45° 26'	5.2-5.5	N/A
NGC 4490	Cocoon Galaxy	CVn	Gx	12h 31m	+41° 39'	9.8	6'
NGC 4631	Whale Galaxy	CVn	Gx	12h 42m	+32° 33'	9.5	13'
RY Dra		Dra	CS	12h 56m	+66° 00'	6.0-8.0	N/A
M 83		Hya	Gx	13h 37m	-29° 52'	7.8	14'
54 Hya		Hya	MS	14h 46m	-25° 27'	5.2	9"
M 68		Hya	GC	12h 39m	-26° 45'	7.3	11'
NGC 3242	Ghost of Jupiter	Hya	PN	10h 25m	-18° 39'	8.6	40"
U Hya		Hya	Var/CS	10h 38m	-13° 23'	4.8-6.5	N/A
R Leo	Peltier's Variable Star	Leo	Var	09h 48m	+11° 26'	4.4-10.5	N/A
Gam Leo	Algieba	Leo	MS	10h 20m	+19° 50'	2.0	5"
54 Leo		Leo	MS	10h 56m	+24° 45'	4.3	6"
M 66		Leo	Gx	11h 20m	+13° 00'	9.7	9'
Iot Leo		Leo	MS	11h 24m	+10° 32'	3.9	2"
38 Lyn		Lyn	MS	09h 19m	+36° 48'	3.8	3"
M 101	Pinwheel Galaxy	UMa	Gx	14h 03m	+54° 21'	8.4	22'
M 40	Winnecke 4	UMa	MS	12h 22m	+58° 05'	9.6	
M 97	Owl Nebula	UMa	PN	11h 15m	+55° 01'	9.7	3'
Zet UMa	Mizar & Alcor	UMa	MS	13h 24m	+54° 56'	2.1	711"
Xi UMa	Alula Australis	UMa	MS	11h 18m	+31° 32'	4.4	2"
VY UMa		UMa	Var/CS	10h 45m	+67° 25'	5.9-6.5	N/A
M 81	Bode's Galaxy	UMa	Gx	09h 56m	+69° 04'	7.8	22'
M 82	Cigar Galaxy	UMa	Gx	09h 56m	+69° 41'	9.0	9'
M 87		Vir	Gx	12h 31m	+12° 23'	9.6	8'
M 104	Sombrero Galaxy	Vir	Gx	12h 40m	-11° 37'	9.1	9'
M 49		Vir	Gx	12h 30m	+08° 00'	9.3	9'
M 60		Vir	Gx	12h 44m	+11° 33'	9.8	7'
The Vir		Vir	MS	13h 10m	-05° 32'	4.4	70"
M 86		Vir	Gx	12h 26m	+12° 57'	9.8	10'
Gam Vir	Porrima	Vir	MS	12h 42m	-01° 27'	2.7	2"
SS Vir		Vir	Var/CS	12h 25m	+00° 48'	6.0-9.6	6.0-9.6

Chart 13

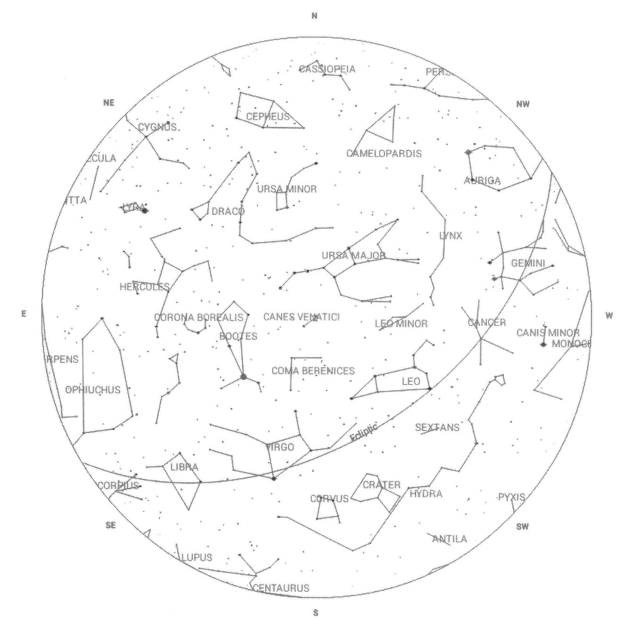

Designation	Name	Con.	Type	R.A.	Dec.	Mag	Size/Sep
Eps Boo	Izar	Boo	MS	14h 45m	+27° 04'	2.4	3"
Xi Boo		Boo	MS	14h 51m	+19° 06'	4.6	6"
Kap Boo	Asellus Tertius	Boo	MS	14h 14m	+51° 47'	4.5	13"
Pi Boo		Boo	MS	14h 41m	+16° 25'	4.5	6"
39 Boo		Boo	MS	14h 50m	+48° 43'	5.7	3"
Struve 1835		Boo	MS	14h 23m	+08° 27'	4.9	6"
Mu Boo	Alkalurops	Boo	MS	15h 24m	+37° 23'	4.3	108"
44 Boo		Boo	MS	15h 04m	+47° 39'	4.8	2"
M 53		Com	GC	13h 13m	+18° 10'	7.7	13'

Designation	Name	Con.	Type	R.A.	Dec.	Mag	Size/Sep
Del Boo		Boo	MS	15h 16m	+33° 19'	3.5	105"
Nu Boo		Boo	MS	15h 31m	+40° 50'	5.0	15'
M 64	Black Eye Galaxy	Com	Gx	12h 57m	+21° 41'	9.3	10'
Melotte 111	Coma Star Cluster	Com	OC	12h 25m	+26° 06'	2.9	120'
24 Com		Com	MS	12h 35m	+18° 23'	5.0	20"
Del Crv	Algorab	Crv	MS	12h 30m	-16° 31'	5.7	24"
Struve 1669		Crv	MS	12h 41m	-13° 01'	5.2	5"
M 51	Whirlpool Galaxy	CVn	Gx	13h 30m	+47° 12'	8.7	10'
M 63	Sunflower Galaxy	CVn	Gx	13h 16m	+42° 02'	9.3	12'
M 106		CVn	Gx	12h 19m	+47° 18'	9.1	17'
M 3		CVn	GC	13h 42m	+28° 23'	6.3	18'
Alp CVn	Cor Caroli	CVn	MS	12h 56m	+38° 19'	2.9	19"
M 94		CVn	Gx	12h 51m	+41° 07'	8.7	10'
2 CVn		CVn	MS	12h 16m	+40° 40'	5.7	11"
NGC 4656	Hook Galaxy	CVn	Gx	12h 44m	+32° 10'	9.7	9'
NGC 4449		CVn	Gx	12h 28m	+44° 06'	9.5	5'
Y CVn	La Superba	CVn	CS	12h 45m	+45° 26'	5.2-5.5	N/A
Zet CrB		CrB	MS	15h 39m	+36° 38'	4.6	6"
R CrB	Fade Out Star	CrB	Var	15h 49m	+28° 09'	5.7-14.8	N/A
NGC 4631	Whale Galaxy	CVn	Gx	12h 42m	+32° 33'	9.5	13'
RY Dra		Dra	CS	12h 56m	+66° 00'	6.0-8.0	N/A
M 83		Hya	Gx	13h 37m	-29° 52'	7.8	14'
54 Hya		Hya	MS	14h 46m	-25° 27'	5.2	9"
M 68		Hya	GC	12h 39m	-26° 45'	7.3	11'
NGC 3242	Ghost of Jupiter	Hya	PN	10h 25m	-18° 39'	8.6	40"
U Hya		Hya	Var/CS	10h 38m	-13° 23'	4.8-6.5	N/A
Gam Leo	Algieba	Leo	MS	10h 20m	+19° 50'	2.0	5"
54 Leo		Leo	MS	10h 56m	+24° 45'	4.3	6"
M 66		Leo	Gx	11h 20m	+13° 00'	9.7	9'
Iot Leo		Leo	MS	11h 24m	+10° 32'	3.9	2"
M 5		Ser	GC	15h 19m	+02° 05'	5.7	23'
Del Ser		Ser	MS	15h 35m	+10° 32'	4.2	4"
M 101	Pinwheel Galaxy	UMa	Gx	14h 03m	+54° 21'	8.4	22'
M 40	Winnecke 4	UMa	MS	12h 22m	+58° 05'	9.6	
M 97	Owl Nebula	UMa	PN	11h 15m	+55° 01'	9.7	3'
Zet UMa	Mizar & Alcor	UMa	MS	13h 24m	+54° 56'	2.1	711"
Xi UMa	Alula Australis	UMa	MS	11h 18m	+31° 32'	4.4	2"
VY UMa		UMa	Var/CS	10h 45m	+67° 25'	5.9-6.5	N/A
M 87		Vir	Gx	12h 31m	+12° 23'	9.6	8'
M 104	Sombrero Galaxy	Vir	Gx	12h 40m	-11° 37'	9.1	9'
M 49		Vir	Gx	12h 30m	+08° 00'	9.3	9'
The Vir		Vir	MS	13h 10m	-05° 32'	4.4	70"
Gam Vir	Porrima	Vir	MS	12h 42m	-01° 27'	2.7	2"
SS Vir		Vir	Var/CS	12h 25m	+00° 48'	6.0-9.6	6.0-9.6

Chart 14

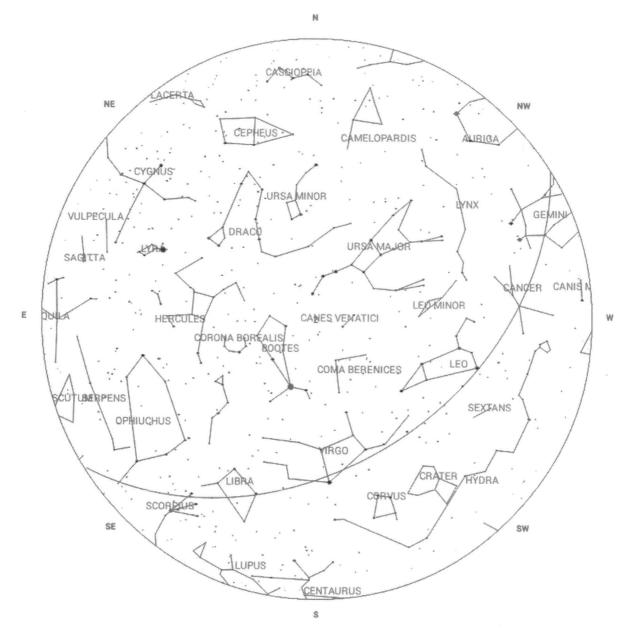

Designation	Name	Con.	Type	R.A.	Dec.	Mag	Size/Sep
Eps Boo	Izar	Boo	MS	14h 45m	+27° 04'	2.4	3"
Xi Boo		Boo	MS	14h 51m	+19° 06'	4.6	6"
Kap Boo	Asellus Tertius	Boo	MS	14h 14m	+51° 47'	4.5	13"
Pi Boo		Boo	MS	14h 41m	+16° 25'	4.5	6"
39 Boo		Boo	MS	14h 50m	+48° 43'	5.7	3"
Struve 1835		Boo	MS	14h 23m	+08° 27'	4.9	6"
Mu Boo	Alkalurops	Boo	MS	15h 24m	+37° 23'	4.3	108"
44 Boo		Boo	MS	15h 04m	+47° 39'	4.8	2"
Del Boo		Boo	MS	15h 16m	+33° 19'	3.5	105"

Designation	Name	Con.	Type	R.A.	Dec.	Mag	Size/Sep
Nu Boo		Boo	MS	15h 31m	+40° 50'	5.0	15'
M 53		Com	GC	13h 13m	+18° 10'	7.7	13'
M 64	Black Eye Galaxy	Com	Gx	12h 57m	+21° 41'	9.3	10'
Melotte 111	Coma Star Cluster	Com	OC	12h 25m	+26° 06'	2.9	120'
24 Com		Com	MS	12h 35m	+18° 23'	5.0	20"
Zet CrB		CrB	MS	15h 39m	+36° 38'	4.6	6"
Sig CrB		CrB	MS	16h 15m	+33° 52'	5.7	7"
R CrB	Fade Out Star	CrB	Var	15h 49m	+28° 09'	5.7-14.8	N/A
T CrB	Blaze Star	CrB	RN	16h 00m	+25° 55'	2.0-10.8	N/A
Del Crv	Algorab	Crv	MS	12h 30m	-16° 31'	5.7	24"
Struve 1669		Crv	MS	12h 41m	-13° 01'	5.2	5"
M 51	Whirlpool Galaxy	CVn	Gx	13h 30m	+47° 12'	8.7	10'
M 63	Sunflower Galaxy	CVn	Gx	13h 16m	+42° 02'	9.3	12'
M 106		CVn	Gx	12h 19m	+47° 18'	9.1	17'
M 3		CVn	GC	13h 42m	+28° 23'	6.3	18'
Alp CVn	Cor Caroli	CVn	MS	12h 56m	+38° 19'	2.9	19"
M 94		CVn	Gx	12h 51m	+41° 07'	8.7	10'
2 CVn		CVn	MS	12h 16m	+40° 40'	5.7	11"
NGC 4449		CVn	Gx	12h 28m	+44° 06'	9.5	5'
Y CVn	La Superba	CVn	CS	12h 45m	+45° 26'	5.2-5.5	N/A
NGC 4631	Whale Galaxy	CVn	Gx	12h 42m	+32° 33'	9.5	13'
RY Dra		Dra	CS	12h 56m	+66° 00'	6.0-8.0	N/A
16/17 Dra		Dra	MS	16h 36m	+52° 55'	5.1	90"
M 13	Keystone Cluster	Her	GC	16h 42m	+36° 27'	5.8	20'
Kap Her	Marfik	Her	MS	16h 41m	+31° 36'	5.0	28"
NGC 6229		Her	GC	16h 47m	+47° 32'	9.4	4'
Iot Leo		Leo	MS	11h 24m	+10° 32'	3.9	2"
Alp Lib	Zuben Elgenubi	Lib	MS	14h 51m	-16° 02'	2.8	230"
NGC 5897	Ghost Globular	Lib	GC	15h 17m	-21° 01'	8.4	11'
Bet Lib	The Emerald Star	Lib	*	15h 14m	-09° 23'	2.6	N/A
Struve 1962		Lib	MS	15h 39m	-08° 47'	5.4	12"
M 5		Ser	GC	15h 19m	+02° 05'	5.7	23'
Del Ser		Ser	MS	15h 35m	+10° 32'	4.2	4"
M 101	Pinwheel Galaxy	UMa	Gx	14h 03m	+54° 21'	8.4	22'
M 40	Winnecke 4	UMa	MS	12h 22m	+58° 05'	9.6	
Zet UMa	Mizar & Alcor	UMa	MS	13h 24m	+54° 56'	2.1	711"
Xi UMa	Alula Australis	UMa	MS	11h 18m	+31° 32'	4.4	2"
M 87		Vir	Gx	12h 31m	+12° 23'	9.6	8'
M 104	Sombrero Galaxy	Vir	Gx	12h 40m	-11° 37'	9.1	9'
M 49		Vir	Gx	12h 30m	+08° 00'	9.3	9'
The Vir		Vir	MS	13h 10m	-05° 32'	4.4	70"
Gam Vir	Porrima	Vir	MS	12h 42m	-01° 27'	2.7	2"
SS Vir		Vir	Var/CS	12h 25m	+00° 48'	6.0-9.6	6.0-9.6

Chart 15

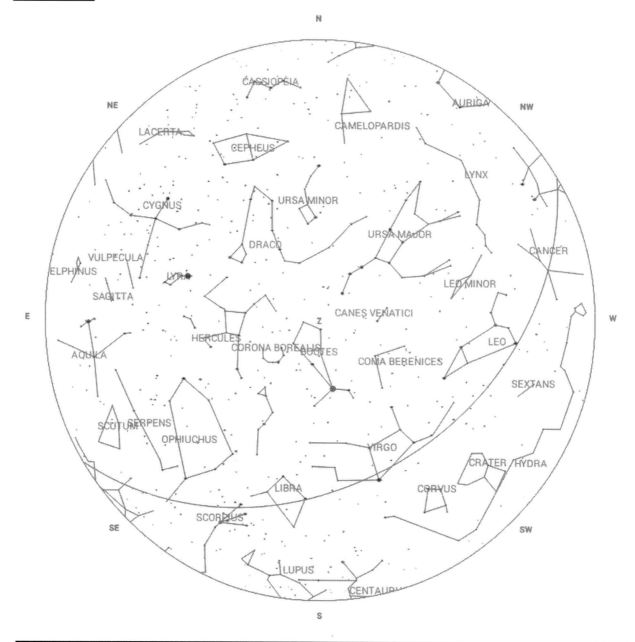

Designation	Name	Con.	Type	R.A.	Dec.	Mag	Size/Sep
Eps Boo	Izar	Boo	MS	14h 45m	+27° 04'	2.4	3"
Xi Boo		Boo	MS	14h 51m	+19° 06'	4.6	6"
Kap Boo	Asellus Tertius	Boo	MS	14h 14m	+51° 47'	4.5	13"
Pi Boo		Boo	MS	14h 41m	+16° 25'	4.5	6"
39 Boo		Boo	MS	14h 50m	+48° 43'	5.7	3"
Struve 1835		Boo	MS	14h 23m	+08° 27'	4.9	6"
Mu Boo	Alkalurops	Boo	MS	15h 24m	+37° 23'	4.3	108"
44 Boo		Boo	MS	15h 04m	+47° 39'	4.8	2"
Del Boo		Boo	MS	15h 16m	+33° 19'	3.5	105"

Designation	Name	Con.	Type	R.A.	Dec.	Mag	Size/Sep
Nu Boo		Boo	MS	15h 31m	+40° 50'	5.0	15'
M 53		Com	GC	13h 13m	+18° 10'	7.7	13'
Melotte 111	Coma Star Cluster	Com	OC	12h 25m	+26° 06'	2.9	120'
24 Com		Com	MS	12h 35m	+18° 23'	5.0	20"
Zet CrB		CrB	MS	15h 39m	+36° 38'	4.6	6"
Sig CrB		CrB	MS	16h 15m	+33° 52'	5.7	7"
R CrB	Fade Out Star	CrB	Var	15h 49m	+28° 09'	5.7-14.8	N/A
T CrB	Blaze Star	CrB	RN	16h 00m	+25° 55'	2.0-10.8	N/A
M 3		CVn	GC	13h 42m	+28° 23'	6.3	18'
Alp CVn	Cor Caroli	CVn	MS	12h 56m	+38° 19'	2.9	19"
2 CVn		CVn	MS	12h 16m	+40° 40'	5.7	11"
Y CVn	La Superba	CVn	CS	12h 45m	+45° 26'	5.2-5.5	N/A
RY Dra		Dra	CS	12h 56m	+66° 00'	6.0-8.0	N/A
16/17 Dra		Dra	MS	16h 36m	+52° 55'	5.1	90"
Nu Dra	Kuma	Dra	MS	17h 32m	+55° 10'	4.9	63"
Mu Dra		Dra	MS	17h 05m	+54° 28'	5.8	2"
Psi Dra		Dra	MS	17h 42m	+72° 09'	4.6	30"
M 13	Keystone Cluster	Her	GC	16h 42m	+36° 27'	5.8	20'
Kap Her	Marfik	Her	MS	16h 41m	+31° 36'	5.0	28"
M 92		Her	GC	17h 17m	+43° 08'	6.5	14'
Alp Her	Rasalgethi	Her	MS	17h 15m	+14° 23'	3.1	5"
Del Her	Sarin	Her	MS	17h 15m	+24° 50'	3.1	14"
Rho Her		Her	MS	17h 24m	+37° 09'	4.2	4"
Alp Lib	Zuben Elgenubi	Lib	MS	14h 51m	-16° 02'	2.8	230"
Bet Lib	The Emerald Star	Lib	*	15h 14m	-09° 23'	2.6	N/A
Struve 1962		Lib	MS	15h 39m	-08° 47'	5.4	12"
IC 4665	Summer Beehive	Oph	OC	17h 46m	+05° 43'	5.3	70'
M 10		Oph	GC	16h 57m	-04° 06'	6.6	20'
M 14		Oph	GC	17h 38m	-03° 15'	7.6	11'
M 12		Oph	GC	16h 47m	-01° 57'	6.1	16'
Rho Oph		Oph	MS	16h 26m	-23° 27'	4.6	3"
M 19		Oph	GC	17h 03m	-26° 16'	6.8	17'
M 62		Oph	GC	17h 01m	-30° 07'	6.4	15'
36 Oph		Oph	MS	17h 15m	-26° 36'	4.3	730"
Omi Oph		Oph	MS	17h 18m	-24° 17'	5.1	10"
61 Oph		Oph	MS	17h 45m	+02° 35'	6.2	21"
M 5		Ser	GC	15h 19m	+02° 05'	5.7	23'
Del Ser		Ser	MS	15h 35m	+10° 32'	4.2	4"
Zet UMa	Mizar & Alcor	UMa	MS	13h 24m	+54° 56'	2.1	711"
The Vir		Vir	MS	13h 10m	-05° 32'	4.4	70"
Gam Vir	Porrima	Vir	MS	12h 42m	-01° 27'	2.7	2"
SS Vir		Vir	Var/CS	12h 25m	+00° 48'	6.0-9.6	6.0-9.6

Chart 16

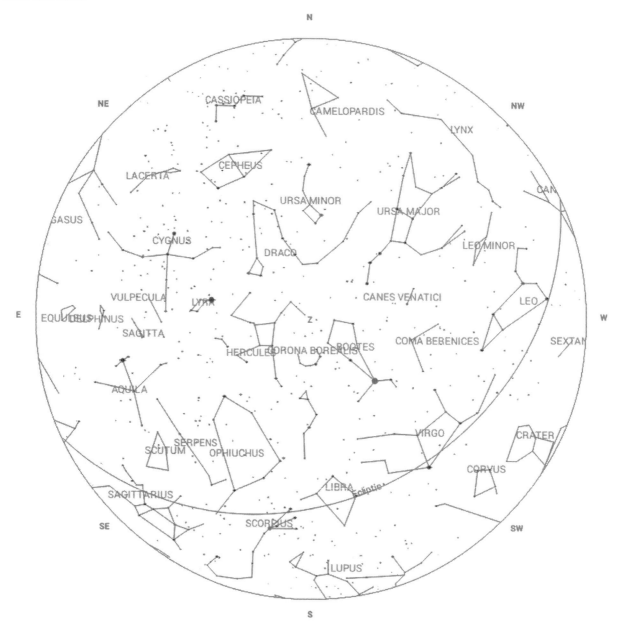

Designation	Name	Con.	Type	R.A.	Dec.	Mag	Size/Sep
Eps Boo	Izar	Boo	MS	14h 45m	+27° 04'	2.4	3"
Xi Boo		Boo	MS	14h 51m	+19° 06'	4.6	6"
Kap Boo	Asellus Tertius	Boo	MS	14h 14m	+51° 47'	4.5	13"
Pi Boo		Boo	MS	14h 41m	+16° 25'	4.5	6"
39 Boo		Boo	MS	14h 50m	+48° 43'	5.7	3"
Struve 1835		Boo	MS	14h 23m	+08° 27'	4.9	6"
Mu Boo	Alkalurops	Boo	MS	15h 24m	+37° 23'	4.3	108"
44 Boo		Boo	MS	15h 04m	+47° 39'	4.8	2"
Del Boo		Boo	MS	15h 16m	+33° 19'	3.5	105"

Designation	Name	Con.	Type	R.A.	Dec.	Mag	Size/Sep
Nu Boo		Boo	MS	15h 31m	+40° 50'	5.0	15'
Zet CrB		CrB	MS	15h 39m	+36° 38'	4.6	6"
Sig CrB		CrB	MS	16h 15m	+33° 52'	5.7	7"
R CrB	Fade Out Star	CrB	Var	15h 49m	+28° 09'	5.7-14.8	N/A
T CrB	Blaze Star	CrB	RN	16h 00m	+25° 55'	2.0-10.8	N/A
M 3		CVn	GC	13h 42m	+28° 23'	6.3	18'
16/17 Dra		Dra	MS	16h 36m	+52° 55'	5.1	90"
Nu Dra	Kuma	Dra	MS	17h 32m	+55° 10'	4.9	63"
Mu Dra		Dra	MS	17h 05m	+54° 28'	5.8	2"
Psi Dra		Dra	MS	17h 42m	+72° 09'	4.6	30"
39 Dra		Dra	MS	18h 24m	+58° 48'	5.0	89"
40/41 Dra		Dra	MS	18h 00m	+80° 00'	5.7	222"
M 13	Keystone Cluster	Her	GC	16h 42m	+36° 27'	5.8	20'
Kap Her	Marfik	Her	MS	16h 41m	+31° 36'	5.0	28"
M 92		Her	GC	17h 17m	+43° 08'	6.5	14'
Alp Her	Rasalgethi	Her	MS	17h 15m	+14° 23'	3.1	5"
Del Her	Sarin	Her	MS	17h 15m	+24° 50'	3.1	14"
Rho Her		Her	MS	17h 24m	+37° 09'	4.2	4"
95 Her		Her	MS	18h 02m	+21° 36'	4.3	6"
100 Her		Her	MS	18h 08m	+26° 06'	5.8	14"
Alp Lib	Zuben Elgenubi	Lib	MS	14h 51m	-16° 02'	2.8	230"
Bet Lib	The Emerald Star	Lib	*	15h 14m	-09° 23'	2.6	N/A
Struve 1962		Lib	MS	15h 39m	-08° 47'	5.4	12"
Eps Lyr	The Double Double	Lyr	MS	18h 44m	+39° 40'	4.7	3"
Bet Lyr	Sheliak	Lyr	MS	18h 50m	+33° 22'	3.2	86"
Del Lyr		Lyr	MS	18h 54m	+36° 58'	4.2	630"
Zet Lyr		Lyr	MS	18h 45m	+37° 36'	4.4	44"
IC 4665	Summer Beehive	Oph	OC	17h 46m	+05° 43'	5.3	70'
M 10		Oph	GC	16h 57m	-04° 06'	6.6	20'
M 12		Oph	GC	16h 47m	-01° 57'	6.1	16'
Rho Oph		Oph	MS	16h 26m	-23° 27'	4.6	3"
M 62		Oph	GC	17h 01m	-30° 07'	6.4	15'
36 Oph		Oph	MS	17h 15m	-26° 36'	4.3	730"
Omi Oph		Oph	MS	17h 18m	-24° 17'	5.1	10"
61 Oph		Oph	MS	17h 45m	+02° 35'	6.2	21"
NGC 6633	Tweedledum Cluster	Oph	OC	18h 27m	+06° 31'	5.6	20'
70 Oph		Oph	MS	18h 06m	+02° 30'	4.2	4"
IC 4756		Ser	OC	18h 39m	+05° 27'	5.4	39'
M 16	Eagle Nebula	Ser	Neb	18h 19m	-13° 49'	6.0	9'
The Ser		Ser	MS	18h 56m	+04° 12'	4.3	22"
M 5		Ser	GC	15h 19m	+02° 05'	5.7	23'
Del Ser		Ser	MS	15h 35m	+10° 32'	4.2	4"
Zet UMa	Mizar & Alcor	UMa	MS	13h 24m	+54° 56'	2.1	711"
The Vir		Vir	MS	13h 10m	-05° 32'	4.4	70"

Chart 17

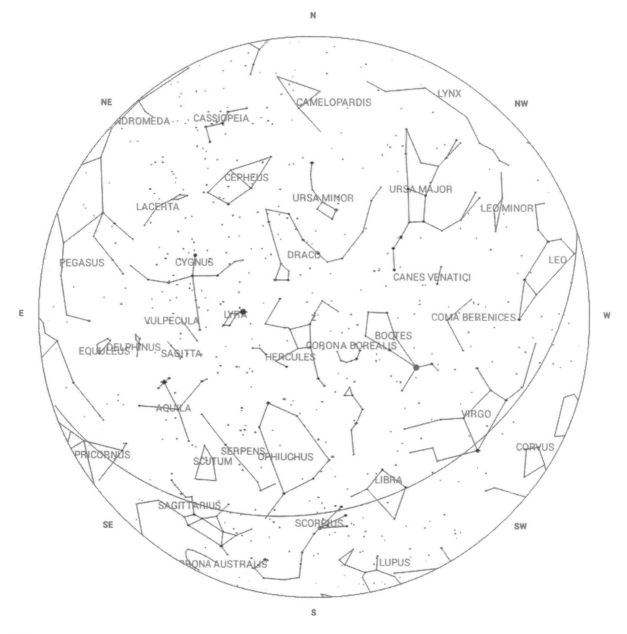

Designation	Name	Con.	Type	R.A.	Dec.	Mag	Size/Sep
15 Aql		Aql	MS	19h 06m	-04° 00'	5.4	39"
57 Aql		Aql	MS	19h 55m	-08° 14'	5.7	36"
Eps Boo	Izar	Boo	MS	14h 45m	+27° 04'	2.4	3"
Xi Boo		Boo	MS	14h 51m	+19° 06'	4.6	6"
Kap Boo	Asellus Tertius	Boo	MS	14h 14m	+51° 47'	4.5	13"
Pi Boo		Boo	MS	14h 41m	+16° 25'	4.5	6"
39 Boo		Boo	MS	14h 50m	+48° 43'	5.7	3"
Struve 1835		Boo	MS	14h 23m	+08° 27'	4.9	6"
Mu Boo	Alkalurops	Boo	MS	15h 24m	+37° 23'	4.3	108"

Designation	Name	Con.	Type	R.A.	Dec.	Mag	Size/Sep
44 Boo		Boo	MS	15h 04m	+47° 39'	4.8	2"
Del Boo		Boo	MS	15h 16m	+33° 19'	3.5	105"
Nu Boo		Boo	MS	15h 31m	+40° 50'	5.0	15'
Zet CrB		CrB	MS	15h 39m	+36° 38'	4.6	6"
Sig CrB		CrB	MS	16h 15m	+33° 52'	5.7	7"
Bet Cyg	Albireo	Cyg	MS	19h 31m	+27° 58'	3.1	34"
Del Cyg		Cyg	MS	19h 45m	+45° 08'	2.9	2"
16/17 Dra		Dra	MS	16h 36m	+52° 55'	5.1	90"
Nu Dra	Kuma	Dra	MS	17h 32m	+55° 10'	4.9	63"
Psi Dra		Dra	MS	17h 42m	+72° 09'	4.6	30"
39 Dra		Dra	MS	18h 24m	+58° 48'	5.0	89"
40/41 Dra		Dra	MS	18h 00m	+80° 00'	5.7	222"
Eps Dra		Dra	MS	19h 48m	+70° 16'	3.8	3"
Kap Her	Marfik	Her	MS	16h 41m	+31° 36'	5.0	28"
Alp Her	Rasalgethi	Her	MS	17h 15m	+14° 23'	3.1	5"
Del Her	Sarin	Her	MS	17h 15m	+24° 50'	3.1	14"
Rho Her		Her	MS	17h 24m	+37° 09'	4.2	4"
95 Her		Her	MS	18h 02m	+21° 36'	4.3	6"
Alp Lib	Zuben Elgenubi	Lib	MS	14h 51m	-16° 02'	2.8	230"
Bet Lib	The Emerald Star	Lib	*	15h 14m	-09° 23'	2.6	N/A
Struve 1962		Lib	MS	15h 39m	-08° 47'	5.4	12"
Eps Lyr	The Double Double	Lyr	MS	18h 44m	+39° 40'	4.7	3"
Bet Lyr	Sheliak	Lyr	MS	18h 50m	+33° 22'	3.2	86"
Del Lyr		Lyr	MS	18h 54m	+36° 58'	4.2	630"
Zet Lyr		Lyr	MS	18h 45m	+37° 36'	4.4	44"
IC 4665	Summer Beehive	Oph	OC	17h 46m	+05° 43'	5.3	70'
Rho Oph		Oph	MS	16h 26m	-23° 27'	4.6	3"
36 Oph		Oph	MS	17h 15m	-26° 36'	4.3	730"
Omi Oph		Oph	MS	17h 18m	-24° 17'	5.1	10"
NGC 6633	Tweedledum Cluster	Oph	OC	18h 27m	+06° 31'	5.6	20'
70 Oph		Oph	MS	18h 06m	+02° 30'	4.2	4"
M 6	Butterfly Cluster	Sco	OC	17h 40m	-32° 15'	4.6	20'
M 7		Sco	OC	17h 54m	-34° 48'	3.3	80'
M 4		Sco	GC	16h 24m	-26° 32'	5.4	36'
Alp Sco	Antares	Sco	MS	16h 29m	-26° 26'	1.0	3"
Bet Sco	Graffias	Sco	MS	16h 05m	-19° 48'	2.6	14"
Nu Sco	Jabbah	Sco	MS	16h 12m	-19° 28'	4.0	2"
Xi Sco		Sco	MS	16h 04m	-11° 22'	4.2	8"
Struve 1999		Sco	MS	16h 04m	-11° 22'	4.2	12"
IC 4756		Ser	OC	18h 39m	+05° 27'	5.4	39'
The Ser		Ser	MS	18h 56m	+04° 12'	4.3	22"
M 5		Ser	GC	15h 19m	+02° 05'	5.7	23'
Del Ser		Ser	MS	15h 35m	+10° 32'	4.2	4"
Collinder 399	Coathanger	Vul	Ast	19h 25m	+20° 11'	4.8	89'

Chart 18

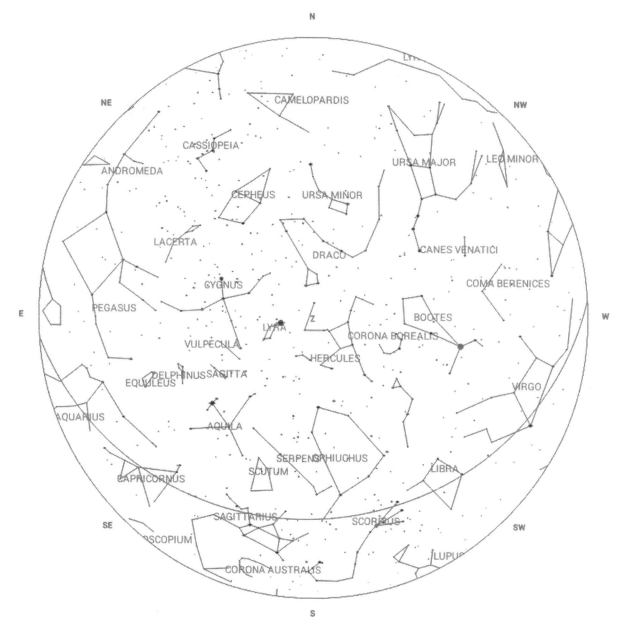

Designation	Name	Con.	Type	R.A.	Dec.	Mag	Size/Sep
15 Aql		Aql	MS	19h 06m	-04° 00'	5.4	39"
57 Aql		Aql	MS	19h 55m	-08° 14'	5.7	36"
Mu Boo	Alkalurops	Boo	MS	15h 24m	+37° 23'	4.3	108"
44 Boo		Boo	MS	15h 04m	+47° 39'	4.8	2"
Del Boo		Boo	MS	15h 16m	+33° 19'	3.5	105"
Nu Boo		Boo	MS	15h 31m	+40° 50'	5.0	15'
Zet CrB		CrB	MS	15h 39m	+36° 38'	4.6	6"
Sig CrB		CrB	MS	16h 15m	+33° 52'	5.7	7"
Bet Cyg	Albireo	Cyg	MS	19h 31m	+27° 58'	3.1	34"

Designation	Name	Con.	Type	R.A.	Dec.	Mag	Size/Sep
Del Cyg		Cyg	MS	19h 45m	+45° 08'	2.9	2"
NGC 7000	North American Nebula	Cyg	Neb	20h 59m	+44° 22'	4.0	120'
Gam Del		Del	MS	20h 47m	+16° 07'	3.9	9"
16/17 Dra		Dra	MS	16h 36m	+52° 55'	5.1	90"
Nu Dra	Kuma	Dra	MS	17h 32m	+55° 10'	4.9	63"
Psi Dra		Dra	MS	17h 42m	+72° 09'	4.6	30"
39 Dra		Dra	MS	18h 24m	+58° 48'	5.0	89"
40/41 Dra		Dra	MS	18h 00m	+80° 00'	5.7	222"
Eps Dra		Dra	MS	19h 48m	+70° 16'	3.8	3"
Eps Equ		Equ	MS	20h 59m	+04° 18'	5.2	11"
Kap Her	Marfik	Her	MS	16h 41m	+31° 36'	5.0	28"
Alp Her	Rasalgethi	Her	MS	17h 15m	+14° 23'	3.1	5"
Del Her	Sarin	Her	MS	17h 15m	+24° 50'	3.1	14"
Rho Her		Her	MS	17h 24m	+37° 09'	4.2	4"
95 Her		Her	MS	18h 02m	+21° 36'	4.3	6"
Eps Lyr	The Double Double	Lyr	MS	18h 44m	+39° 40'	4.7	3"
Bet Lyr	Sheliak	Lyr	MS	18h 50m	+33° 22'	3.2	86"
Del Lyr		Lyr	MS	18h 54m	+36° 58'	4.2	630"
Zet Lyr		Lyr	MS	18h 45m	+37° 36'	4.4	44"
IC 4665	Summer Beehive	Oph	OC	17h 46m	+05° 43'	5.3	70'
Rho Oph		Oph	MS	16h 26m	-23° 27'	4.6	3"
36 Oph		Oph	MS	17h 15m	-26° 36'	4.3	730"
Omi Oph		Oph	MS	17h 18m	-24° 17'	5.1	10"
NGC 6633	Tweedledum Cluster	Oph	OC	18h 27m	+06° 31'	5.6	20'
70 Oph		Oph	MS	18h 06m	+02° 30'	4.2	4"
M 6	Butterfly Cluster	Sco	OC	17h 40m	-32° 15'	4.6	20'
M 7		Sco	OC	17h 54m	-34° 48'	3.3	80'
M 4		Sco	GC	16h 24m	-26° 32'	5.4	36'
Alp Sco	Antares	Sco	MS	16h 29m	-26° 26'	1.0	3"
Bet Sco	Graffias	Sco	MS	16h 05m	-19° 48'	2.6	14"
Nu Sco	Jabbah	Sco	MS	16h 12m	-19° 28'	4.0	2"
Xi Sco		Sco	MS	16h 04m	-11° 22'	4.2	8"
Struve 1999		Sco	MS	16h 04m	-11° 22'	4.2	12"
IC 4756		Ser	OC	18h 39m	+05° 27'	5.4	39'
The Ser		Ser	MS	18h 56m	+04° 12'	4.3	22"
M 5		Ser	GC	15h 19m	+02° 05'	5.7	23'
Del Ser		Ser	MS	15h 35m	+10° 32'	4.2	4"
The Sge		Sge	MS	20h 10m	+20° 55'	4.6	84"
M 24	Sagittarius Star Cloud	Sgr	OC	18h 18m	-18° 24'	3.1	90'
M 22		Sgr	GC	18h 36m	-23° 54'	5.2	32'
M 8	Lagoon Nebula	Sgr	Neb	18h 04m	-24° 23'	5.0	17'
Collinder 399	Coathanger	Vul	Ast	19h 25m	+20° 11'	4.8	89'
NGC 6885		Vul	OC	20h 12m	+26° 29'	5.7	20'

Chart 19

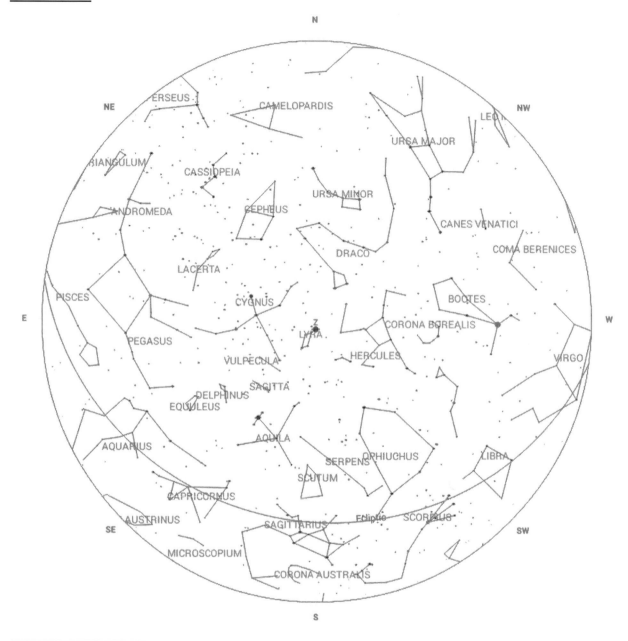

Designation	Name	Con.	Type	R.A.	Dec.	Mag	Size/Sep
15 Aql		Aql	MS	19h 06m	-04° 00'	5.4	39"
57 Aql		Aql	MS	19h 55m	-08° 14'	5.7	36"
Bet Cyg	Albireo	Cyg	MS	19h 31m	+27° 58'	3.1	34"
Del Cyg		Cyg	MS	19h 45m	+45° 08'	2.9	2"
16 Cyg		Cyg	MS	19h 42m	+50° 32'	5.9	39"
61 Cyg		Cyg	MS	21h 07m	+38° 44'	5.2	29"
M 39		Cyg	OC	21h 32m	+48° 26'	5.3	29'
Mu Cyg		Cyg	MS	21h 44m	+28° 45'	4.5	200"

Designation	Name	Con.	Type	R.A.	Dec.	Mag	Size/Sep
V460		Cyg	Var/CS	21h 42m	+35° 31'	5.6-7.0	N/A
NGC 7000	North American Nebula	Cyg	Neb	20h 59m	+44° 22'	4.0	120'
	Northern Coalsack	Cyg	DN	20h 41m	+43° 00'	6.0	60'
Gam Del		Del	MS	20h 47m	+16° 07'	3.9	9"
16/17 Dra		Dra	MS	16h 36m	+52° 55'	5.1	90"
Nu Dra	Kuma	Dra	MS	17h 32m	+55° 10'	4.9	63"
Mu Dra		Dra	MS	17h 05m	+54° 28'	5.8	2"
Psi Dra		Dra	MS	17h 42m	+72° 09'	4.6	30"
39 Dra		Dra	MS	18h 24m	+58° 48'	5.0	89"
40/41 Dra		Dra	MS	18h 00m	+80° 00'	5.7	222"
Eps Dra		Dra	MS	19h 48m	+70° 16'	3.8	3"
Eps Equ		Equ	MS	20h 59m	+04° 18'	5.2	11"
M 13	Keystone Cluster	Her	GC	16h 42m	+36° 27'	5.8	20'
Kap Her	Marfik	Her	MS	16h 41m	+31° 36'	5.0	28"
Alp Her	Rasalgethi	Her	MS	17h 15m	+14° 23'	3.1	5"
Del Her	Sarin	Her	MS	17h 15m	+24° 50'	3.1	14"
Rho Her		Her	MS	17h 24m	+37° 09'	4.2	4"
95 Her		Her	MS	18h 02m	+21° 36'	4.3	6"
100 Her		Her	MS	18h 08m	+26° 06'	5.8	14"
Eps Lyr	The Double Double	Lyr	MS	18h 44m	+39° 40'	4.7	3"
Bet Lyr	Sheliak	Lyr	MS	18h 50m	+33° 22'	3.2	86"
Del Lyr		Lyr	MS	18h 54m	+36° 58'	4.2	630"
Zet Lyr		Lyr	MS	18h 45m	+37° 36'	4.4	44"
IC 4665	Summer Beehive	Oph	OC	17h 46m	+05° 43'	5.3	70'
M 12		Oph	GC	16h 47m	-01° 57'	6.1	16'
Rho Oph		Oph	MS	16h 26m	-23° 27'	4.6	3"
36 Oph		Oph	MS	17h 15m	-26° 36'	4.3	730"
Omi Oph		Oph	MS	17h 18m	-24° 17'	5.1	10"
61 Oph		Oph	MS	17h 45m	+02° 35'	6.2	21"
NGC 6633	Tweedledum Cluster	Oph	OC	18h 27m	+06° 31'	5.6	20'
70 Oph		Oph	MS	18h 06m	+02° 30'	4.2	4"
IC 4756		Ser	OC	18h 39m	+05° 27'	5.4	39'
M 16	Eagle Nebula	Ser	Neb	18h 19m	-13° 49'	6.0	9'
The Ser		Ser	MS	18h 56m	+04° 12'	4.3	22"
The Sge		Sge	MS	20h 10m	+20° 55'	4.6	84"
15 Sge		Sge	MS	20h 04m	+17° 04'	5.8	204"
M 17	Swan Nebula	Sgr	Neb	18h 21m	-16° 11'	6.0	11'
M 25		Sgr	OC	18h 32m	-19° 07'	6.2	29'
M 24	Sagittarius Star Cloud	Sgr	OC	18h 18m	-18° 24'	3.1	90'
M 22		Sgr	GC	18h 36m	-23° 54'	5.2	32'
M 8	Lagoon Nebula	Sgr	Neb	18h 04m	-24° 23'	5.0	17'
M 23		Sgr	OC	17h 57m	-18° 59'	5.9	29'
Collinder 399	Coathanger	Vul	Ast	19h 25m	+20° 11'	4.8	89'
NGC 6885		Vul	OC	20h 12m	+26° 29'	5.7	20'

Chart 20

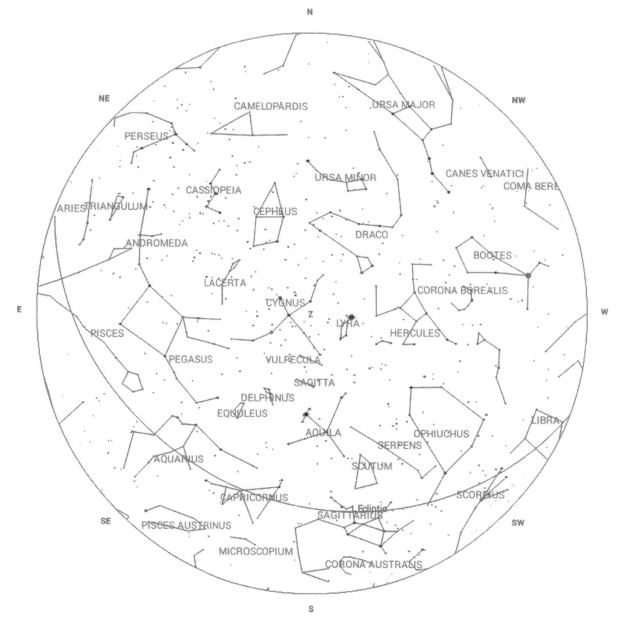

Designation	Name	Con.	Type	R.A.	Dec.	Mag	Size/Sep
15 Aql		Aql	MS	19h 06m	-04° 00'	5.4	39"
57 Aql		Aql	MS	19h 55m	-08° 14'	5.7	36"
Alp Cap	Al Giedi	Cap	MS	20h 18m	-12° 32'	3.6	45"
Bet Cap	Dabih	Cap	MS	20h 21m	-14° 47'	3.1	205"
Omi Cap		Cap	MS	20h 30m	-18° 35'	5.9	22"
IC 1396	Misty Clover Cluster	Cep	OC	21h 39m	+57° 30'	5.1	89'
Bet Cep	Alfirk	Cep	MS	21h 29m	+70° 34'	3.2	13"
Del Cep		Cep	MS/Var	22h 29m	+58° 25'	3.5-4.4	41"
Struve 2816		Cep	MS	21h 39m	+57° 29'	5.7	20"

Designation	Name	Con.	Type	R.A.	Dec.	Mag	Size/Sep
Mu Cep	Herschel's Garnet Star	Cep	Var/CS	21h 44m	+58° 47'	3.4-5.1	N/A
Struve 2840		Cep	MS	21h 52m	+55° 48'	5.7	18"
Xi Cep	Alkurhah	Cep	MS	22h 04m	+64° 38'	4.3	8"
Bet Cyg	Albireo	Cyg	MS	19h 31m	+27° 58'	3.1	34"
Del Cyg		Cyg	MS	19h 45m	+45° 08'	2.9	2"
16 Cyg		Cyg	MS	19h 42m	+50° 32'	5.9	39"
61 Cyg		Cyg	MS	21h 07m	+38° 44'	5.2	29"
M 39		Cyg	OC	21h 32m	+48° 26'	5.3	29'
Mu Cyg		Cyg	MS	21h 44m	+28° 45'	4.5	200"
NGC 7000	North American Nebula	Cyg	Neb	20h 59m	+44° 22'	4.0	120'
Gam Del		Del	MS	20h 47m	+16° 07'	3.9	9"
Nu Dra	Kuma	Dra	MS	17h 32m	+55° 10'	4.9	63"
Mu Dra		Dra	MS	17h 05m	+54° 28'	5.8	2"
Psi Dra		Dra	MS	17h 42m	+72° 09'	4.6	30"
39 Dra		Dra	MS	18h 24m	+58° 48'	5.0	89"
40/41 Dra		Dra	MS	18h 00m	+80° 00'	5.7	222"
Eps Dra		Dra	MS	19h 48m	+70° 16'	3.8	3"
Eps Equ		Equ	MS	20h 59m	+04° 18'	5.2	11"
Alp Her	Rasalgethi	Her	MS	17h 15m	+14° 23'	3.1	5"
Del Her	Sarin	Her	MS	17h 15m	+24° 50'	3.1	14"
Rho Her		Her	MS	17h 24m	+37° 09'	4.2	4"
95 Her		Her	MS	18h 02m	+21° 36'	4.3	6"
100 Her		Her	MS	18h 08m	+26° 06'	5.8	14"
8 Lac		Lac	MS	22h 36m	+39° 38'	5.7	82"
Eps Lyr	The Double Double	Lyr	MS	18h 44m	+39° 40'	4.7	3"
Bet Lyr	Sheliak	Lyr	MS	18h 50m	+33° 22'	3.2	86"
Del Lyr		Lyr	MS	18h 54m	+36° 58'	4.2	630"
Zet Lyr		Lyr	MS	18h 45m	+37° 36'	4.4	44"
IC 4665	Summer Beehive	Oph	OC	17h 46m	+05° 43'	5.3	70'
36 Oph		Oph	MS	17h 15m	-26° 36'	4.3	730"
Omi Oph		Oph	MS	17h 18m	-24° 17'	5.1	10"
NGC 6633	Tweedledum Cluster	Oph	OC	18h 27m	+06° 31'	5.6	20'
70 Oph		Oph	MS	18h 06m	+02° 30'	4.2	4"
Eps Peg	Enif	Peg	MS	21h 44m	+09° 52'	2.1	143"
IC 4756		Ser	OC	18h 39m	+05° 27'	5.4	39'
The Ser		Ser	MS	18h 56m	+04° 12'	4.3	22"
The Sge		Sge	MS	20h 10m	+20° 55'	4.6	84"
15 Sge		Sge	MS	20h 04m	+17° 04'	5.8	204"
M 24	Sagittarius Star Cloud	Sgr	OC	18h 18m	-18° 24'	3.1	90'
M 22		Sgr	GC	18h 36m	-23° 54'	5.2	32'
M 8	Lagoon Nebula	Sgr	Neb	18h 04m	-24° 23'	5.0	17'
M 23		Sgr	OC	17h 57m	-18° 59'	5.9	29'
Collinder 399	Coathanger	Vul	Ast	19h 25m	+20° 11'	4.8	89'
NGC 6885		Vul	OC	20h 12m	+26° 29'	5.7	20'

Chart 21

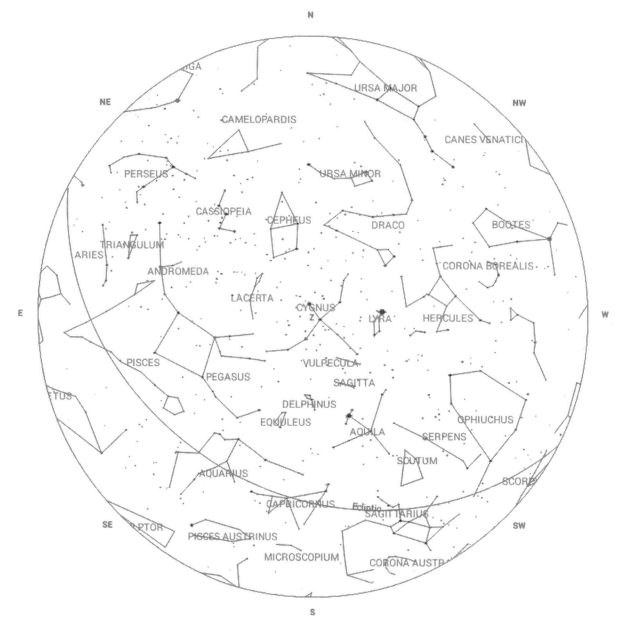

Designation	Name	Con.	Type	R.A.	Dec.	Mag	Size/Sep
Struve 2404		Aql	MS	18h 51m	+10° 59'	6.4	4"
15 Aql		Aql	MS	19h 06m	-04° 00'	5.4	39"
57 Aql		Aql	MS	19h 55m	-08° 14'	5.7	36"
V Aql		Aql	Var	19h 04m	-05° 41'	6.6-8.4	N/A
107 Aqr		Aqr	MS	23h 47m	-18° 35'	5.3	7"
94 Aqr		Aqr	MS	23h 19m	-13° 28'	5.2	13"
NGC 7293	Helix Nebula	Aqr	PN	22h 30m	-20° 50'	6.3	16'
Zet Aqr		Aqr	MS	22h 29m	-00° 01'	3.7	2"
41 Aqr		Aqr	MS	22h 14m	-21° 04'	5.3	5"

Designation	Name	Con.	Type	R.A.	Dec.	Mag	Size/Sep
53 Aqr		Aqr	MS	22h 27m	-16° 45'	5.6	3"
M 2		Aqr	GC	21h 33m	-00° 49'	6.6	16'
Alp Cap	Al Giedi	Cap	MS	20h 18m	-12° 32'	3.6	45"
Bet Cap	Dabih	Cap	MS	20h 21m	-14° 47'	3.1	205"
RT Cap		Cap	Var/CS	20h 17m	-21° 19'	6.5-8.1	N/A
Omi Cap		Cap	MS	20h 30m	-18° 35'	5.9	22"
Sig Cas		Cas	MS	23h 59m	+55° 45'	4.9	3"
IC 1396	Misty Clover Cluster	Cep	OC	21h 39m	+57° 30'	5.1	89'
Bet Cep	Alfirk	Cep	MS	21h 29m	+70° 34'	3.2	13"
Del Cep		Cep	MS/Var	22h 29m	+58° 25'	3.5-4.4	41"
Struve 2816		Cep	MS	21h 39m	+57° 29'	5.7	20"
Mu Cep	Herschel's Garnet Star	Cep	Var/CS	21h 44m	+58° 47'	3.4-5.1	N/A
Struve 2840		Cep	MS	21h 52m	+55° 48'	5.7	18"
Xi Cep	Alkurhah	Cep	MS	22h 04m	+64° 38'	4.3	8"
Omi Cep		Cep	MS	23h 19m	+68° 07'	4.8	3"
Bet Cyg	Albireo	Cyg	MS	19h 31m	+27° 58'	3.1	34"
Del Cyg		Cyg	MS	19h 45m	+45° 08'	2.9	2"
16 Cyg		Cyg	MS	19h 42m	+50° 32'	5.9	39"
61 Cyg		Cyg	MS	21h 07m	+38° 44'	5.2	29"
M 39		Cyg	OC	21h 32m	+48° 26'	5.3	29'
Mu Cyg		Cyg	MS	21h 44m	+28° 45'	4.5	200"
V460		Cyg	Var/CS	21h 42m	+35° 31'	5.6-7.0	N/A
NGC 7000	North American Nebula	Cyg	Neb	20h 59m	+44° 22'	4.0	120'
	Northern Coalsack	Cyg	DN	20h 41m	+43° 00'	6.0	60'
Gam Del		Del	MS	20h 47m	+16° 07'	3.9	9"
39 Dra		Dra	MS	18h 24m	+58° 48'	5.0	89"
40/41 Dra		Dra	MS	18h 00m	+80° 00'	5.7	222"
Eps Dra		Dra	MS	19h 48m	+70° 16'	3.8	3"
UX Dra		Dra	Var/CS	19h 22m	+76° 34'	5.9-7.1	N/A
Eps Equ		Equ	MS	20h 59m	+04° 18'	5.2	11"
95 Her		Her	MS	18h 02m	+21° 36'	4.3	6"
100 Her		Her	MS	18h 08m	+26° 06'	5.8	14"
8 Lac		Lac	MS	22h 36m	+39° 38'	5.7	82"
Eps Lyr	The Double Double	Lyr	MS	18h 44m	+39° 40'	4.7	3"
Bet Lyr	Sheliak	Lyr	MS	18h 50m	+33° 22'	3.2	86"
Del Lyr		Lyr	MS	18h 54m	+36° 58'	4.2	630"
Zet Lyr		Lyr	MS	18h 45m	+37° 36'	4.4	44"
M 15		Peg	GC	21h 30m	+12° 10'	6.3	18'
Eps Peg	Enif	Peg	MS	21h 44m	+09° 52'	2.1	143"
M 11	Wild Duck Cluster	Sct	OC	18h 51m	-06° 16'	6.1	32'
The Sge		Sge	MS	20h 10m	+20° 55'	4.6	84"
15 Sge		Sge	MS	20h 04m	+17° 04'	5.8	204"
Collinder 399	Coathanger	Vul	Ast	19h 25m	+20° 11'	4.8	89'
NGC 6885		Vul	OC	20h 12m	+26° 29'	5.7	20'

Chart 22

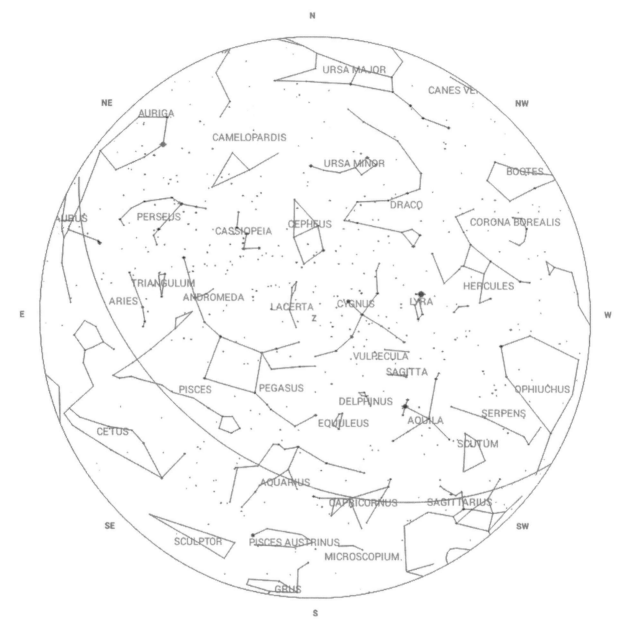

Designation	Name	Con.	Type	R.A.	Dec.	Mag	Size/Sep
M 31	Andromeda Galaxy	And	Gx	00h 43m	+41° 16'	4.3	156'
Pi And		And	MS	00h 38m	+33° 49'	4.4	36"
15 Aql		Aql	MS	19h 06m	-04° 00'	5.4	39"
57 Aql		Aql	MS	19h 55m	-08° 14'	5.7	36"
107 Aqr		Aqr	MS	23h 47m	-18° 35'	5.3	7"
94 Aqr		Aqr	MS	23h 19m	-13° 28'	5.2	13"
NGC 7293	Helix Nebula	Aqr	PN	22h 30m	-20° 50'	6.3	16'
Zet Aqr		Aqr	MS	22h 29m	-00° 01'	3.7	2"

Designation	Name	Con.	Type	R.A.	Dec.	Mag	Size/Sep
41 Aqr		Aqr	MS	22h 14m	-21° 04'	5.3	5"
53 Aqr		Aqr	MS	22h 27m	-16° 45'	5.6	3"
M 2		Aqr	GC	21h 33m	-00° 49'	6.6	16'
M 30		Cap	GC	21h 40m	-23° 11'	6.9	12'
Alp Cap	Al Giedi	Cap	MS	20h 18m	-12° 32'	3.6	45"
Bet Cap	Dabih	Cap	MS	20h 21m	-14° 47'	3.1	205"
Omi Cap		Cap	MS	20h 30m	-18° 35'	5.9	22"
Struve 3053		Cas	MS	00h 03m	+66° 06'	5.9	15"
Eta Cas	Achird	Cas	MS	00h 50m	+57° 54'	3.6	13"
Sig Cas		Cas	MS	23h 59m	+55° 45'	4.9	3"
IC 1396	Misty Clover Cluster	Cep	OC	21h 39m	+57° 30'	5.1	89'
Bet Cep	Alfirk	Cep	MS	21h 29m	+70° 34'	3.2	13"
Del Cep		Cep	MS/Var	22h 29m	+58° 25'	3.5-4.4	41"
Struve 2816		Cep	MS	21h 39m	+57° 29'	5.7	20"
Mu Cep	Herschel's Garnet Star	Cep	Var/CS	21h 44m	+58° 47'	3.4-5.1	N/A
Struve 2840		Cep	MS	21h 52m	+55° 48'	5.7	18"
Xi Cep	Alkurhah	Cep	MS	22h 04m	+64° 38'	4.3	8"
Omi Cep		Cep	MS	23h 19m	+68° 07'	4.8	3"
Bet Cyg	Albireo	Cyg	MS	19h 31m	+27° 58'	3.1	34"
Del Cyg		Cyg	MS	19h 45m	+45° 08'	2.9	2"
16 Cyg		Cyg	MS	19h 42m	+50° 32'	5.9	39"
NGC 6960	Veil Nebula (West)	Cyg	SNR	20h 46m	+30° 43'	7.0	63'
61 Cyg		Cyg	MS	21h 07m	+38° 44'	5.2	29"
M 39		Cyg	OC	21h 32m	+48° 26'	5.3	29'
Mu Cyg		Cyg	MS	21h 44m	+28° 45'	4.5	200"
V460		Cyg	Var/CS	21h 42m	+35° 31'	5.6-7.0	N/A
NGC 6992	Veil Nebula (East)	Cyg	SNR	20h 56m	+31° 43'	7.0	60'
NGC 7000	North American Nebula	Cyg	Neb	20h 59m	+44° 22'	4.0	120'
	Northern Coalsack	Cyg	DN	20h 41m	+43° 00'	6.0	60'
Gam Del		Del	MS	20h 47m	+16° 07'	3.9	9"
Eps Dra		Dra	MS	19h 48m	+70° 16'	3.8	3"
Eps Equ		Equ	MS	20h 59m	+04° 18'	5.2	11"
NGC 7243		Lac	OC	22h 15m	+49° 54'	6.7	29'
8 Lac		Lac	MS	22h 36m	+39° 38'	5.7	82"
Struve 2470/2474	The Double Double's Double	Lyr	MS	19h 09m	+34° 41'	6.7	16"
M 15		Peg	GC	21h 30m	+12° 10'	6.3	18'
Eps Peg	Enif	Peg	MS	21h 44m	+09° 52'	2.1	143"
TX Psc		Psc	Var/CS	23h 46m	+03° 29'	4.5-5.3	N/A
55 Psc		Psc	MS	00h 40m	+21° 26'	5.4	6"
65 Psc		Psc	MS	00h 50m	+27° 43'	7.0	4"
The Sge		Sge	MS	20h 10m	+20° 55'	4.6	84"
15 Sge		Sge	MS	20h 04m	+17° 04'	5.8	204"
Collinder 399	Coathanger	Vul	Ast	19h 25m	+20° 11'	4.8	89'
NGC 6885		Vul	OC	20h 12m	+26° 29'	5.7	20'

Chart 23

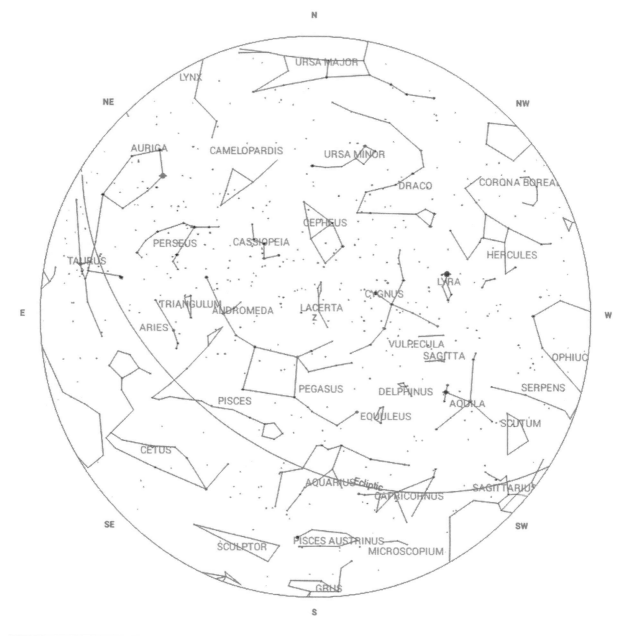

Designation	Name	Con.	Type	R.A.	Dec.	Mag	Size/Sep
M 31	Andromeda Galaxy	And	Gx	00h 43m	+41° 16'	4.3	156'
NGC 752	Golf Ball Cluster	And	OC	01h 58m	+37° 47'	6.6	75'
Pi And		And	MS	00h 38m	+33° 49'	4.4	36"
107 Aqr		Aqr	MS	23h 47m	-18° 35'	5.3	7"
94 Aqr		Aqr	MS	23h 19m	-13° 28'	5.2	13"
NGC 7293	Helix Nebula	Aqr	PN	22h 30m	-20° 50'	6.3	16'
Zet Aqr		Aqr	MS	22h 29m	-00° 01'	3.7	2"
41 Aqr		Aqr	MS	22h 14m	-21° 04'	5.3	5"

Designation	Name	Con.	Type	R.A.	Dec.	Mag	Size/Sep
53 Aqr		Aqr	MS	22h 27m	-16° 45'	5.6	3"
M 2		Aqr	GC	21h 33m	-00° 49'	6.6	16'
Lam Ari		Ari	MS	01h 59m	+23° 41'	4.8	37"
Gam Ari	Mesarthim	Ari	MS	01h 54m	+19° 22'	4.6	8"
M 30		Cap	GC	21h 40m	-23° 11'	6.9	12'
Alp Cap	Al Giedi	Cap	MS	20h 18m	-12° 32'	3.6	45"
Bet Cap	Dabih	Cap	MS	20h 21m	-14° 47'	3.1	205"
Omi Cap		Cap	MS	20h 30m	-18° 35'	5.9	22"
NGC 457	Owl Cluster	Cas	OC	01h 20m	+58° 17'	5.1	20'
Struve 163		Cas	MS	01h 51m	+64° 51'	6.5	35"
M 103		Cas	OC	01h 33m	+60° 39'	6.9	5'
Struve 3053		Cas	MS	00h 03m	+66° 06'	5.9	15"
Eta Cas	Achird	Cas	MS	00h 50m	+57° 54'	3.6	13"
Sig Cas		Cas	MS	23h 59m	+55° 45'	4.9	3"
NGC 663		Cas	OC	01h 46m	+61° 14'	6.4	14'
IC 1396	Misty Clover Cluster	Cep	OC	21h 39m	+57° 30'	5.1	89'
Bet Cep	Alfirk	Cep	MS	21h 29m	+70° 34'	3.2	13"
Del Cep		Cep	MS/Var	22h 29m	+58° 25'	3.5-4.4	41"
Struve 2816		Cep	MS	21h 39m	+57° 29'	5.7	20"
Mu Cep	Herschel's Garnet Star	Cep	Var/CS	21h 44m	+58° 47'	3.4-5.1	N/A
Struve 2840		Cep	MS	21h 52m	+55° 48'	5.7	18"
Xi Cep	Alkurhah	Cep	MS	22h 04m	+64° 38'	4.3	8"
Omi Cep		Cep	MS	23h 19m	+68° 07'	4.8	3"
NGC 6960	Veil Nebula (West)	Cyg	SNR	20h 46m	+30° 43'	7.0	63'
61 Cyg		Cyg	MS	21h 07m	+38° 44'	5.2	29"
M 39		Cyg	OC	21h 32m	+48° 26'	5.3	29'
Mu Cyg		Cyg	MS	21h 44m	+28° 45'	4.5	200"
NGC 6992	Veil Nebula (East)	Cyg	SNR	20h 56m	+31° 43'	7.0	60'
NGC 7000	North American Nebula	Cyg	Neb	20h 59m	+44° 22'	4.0	120'
	Northern Coalsack	Cyg	DN	20h 41m	+43° 00'	6.0	60'
Gam Del		Del	MS	20h 47m	+16° 07'	3.9	9"
Eps Equ		Equ	MS	20h 59m	+04° 18'	5.2	11"
NGC 7243		Lac	OC	22h 15m	+49° 54'	6.7	29'
8 Lac		Lac	MS	22h 36m	+39° 38'	5.7	82"
M 15		Peg	GC	21h 30m	+12° 10'	6.3	18'
Eps Peg	Enif	Peg	MS	21h 44m	+09° 52'	2.1	143"
55 Psc		Psc	MS	00h 40m	+21° 26'	5.4	6"
65 Psc		Psc	MS	00h 50m	+27° 43'	7.0	4"
Psi1 Psc		Psc	MS	01h 06m	+21° 28'	5.3	30"
Zet Psc		Psc	MS	01h 14m	+07° 35'	5.2	23"
The Sge		Sge	MS	20h 10m	+20° 55'	4.6	84"
15 Sge		Sge	MS	20h 04m	+17° 04'	5.8	204"
M 33	Triangulum Galaxy	Tri	Gx	01h 34m	+30° 40'	6.4	62'
NGC 6885		Vul	OC	20h 12m	+26° 29'	5.7	20'

Chart 24

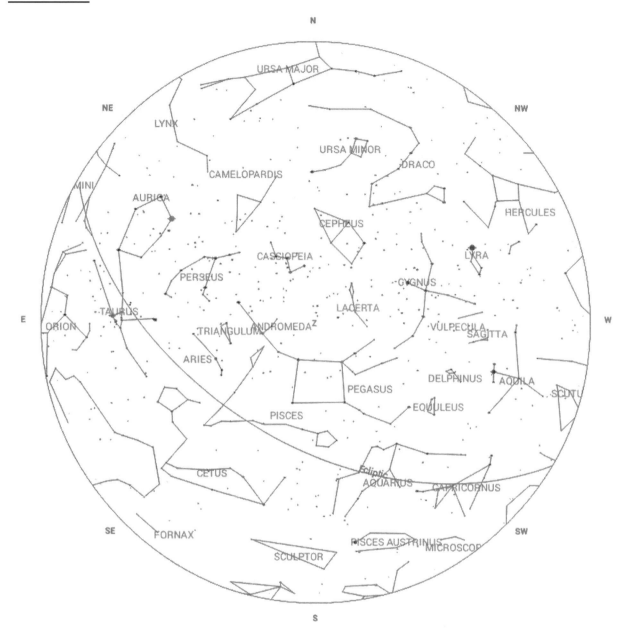

Designation	Name	Con.	Type	R.A.	Dec.	Mag	Size/Sep
Gam And	Almach	And	MS	02h 04m	+42° 20'	2.1	10"
M 31	Andromeda Galaxy	And	Gx	00h 43m	+41° 16'	4.3	156'
NGC 752	Golf Ball Cluster	And	OC	01h 58m	+37° 47'	6.6	75'
Pi And		And	MS	00h 38m	+33° 49'	4.4	36"
107 Aqr		Aqr	MS	23h 47m	-18° 35'	5.3	7"
94 Aqr		Aqr	MS	23h 19m	-13° 28'	5.2	13"
NGC 7293	Helix Nebula	Aqr	PN	22h 30m	-20° 50'	6.3	16'
Zet Aqr		Aqr	MS	22h 29m	-00° 01'	3.7	2"

Designation	Name	Con.	Type	R.A.	Dec.	Mag	Size/Sep
41 Aqr		Aqr	MS	22h 14m	-21° 04'	5.3	5"
53 Aqr		Aqr	MS	22h 27m	-16° 45'	5.6	3"
M 2		Aqr	GC	21h 33m	-00° 49'	6.6	16'
Lam Ari		Ari	MS	01h 59m	+23° 41'	4.8	37"
Gam Ari	Mesarthim	Ari	MS	01h 54m	+19° 22'	4.6	8"
30 Ari		Ari	MS	02h 37m	+24° 39'	6.5	39"
M 30		Cap	GC	21h 40m	-23° 11'	6.9	12'
NGC 457	Owl Cluster	Cas	OC	01h 20m	+58° 17'	5.1	20'
Struve 163		Cas	MS	01h 51m	+64° 51'	6.5	35"
M 103		Cas	OC	01h 33m	+60° 39'	6.9	5'
Struve 3053		Cas	MS	00h 03m	+66° 06'	5.9	15"
Eta Cas	Achird	Cas	MS	00h 50m	+57° 54'	3.6	13"
Sig Cas		Cas	MS	23h 59m	+55° 45'	4.9	3"
Iot Cas		Cas	MS	02h 29m	+67° 24'	4.5	7"
NGC 659	Ying Yang Cluster	Cas	OC	01h 44m	+60° 40'	7.2	5'
NGC 663		Cas	OC	01h 46m	+61° 14'	6.4	14'
IC 1396	Misty Clover Cluster	Cep	OC	21h 39m	+57° 30'	5.1	89'
Bet Cep	Alfirk	Cep	MS	21h 29m	+70° 34'	3.2	13"
Del Cep		Cep	MS/Var	22h 29m	+58° 25'	3.5-4.4	41"
Struve 2816		Cep	MS	21h 39m	+57° 29'	5.7	20"
Mu Cep	Herschel's Garnet Star	Cep	Var/CS	21h 44m	+58° 47'	3.4-5.1	N/A
Struve 2840		Cep	MS	21h 52m	+55° 48'	5.7	18"
Xi Cep	Alkurhah	Cep	MS	22h 04m	+64° 38'	4.3	8"
Omi Cep		Cep	MS	23h 19m	+68° 07'	4.8	3"
Gam Cet	Kaffajidhma	Cet	MS	02h 43m	+03° 14'	3.5	3"
Omi Cet	Mira	Cet	Var	02h 19m	-02° 59'	2.0-10.1	N/A
61 Cyg		Cyg	MS	21h 07m	+38° 44'	5.2	29"
M 39		Cyg	OC	21h 32m	+48° 26'	5.3	29'
Mu Cyg		Cyg	MS	21h 44m	+28° 45'	4.5	200"
V460		Cyg	Var/CS	21h 42m	+35° 31'	5.6-7.0	N/A
NGC 7243		Lac	OC	22h 15m	+49° 54'	6.7	29'
8 Lac		Lac	MS	22h 36m	+39° 38'	5.7	82"
M 34		Per	OC	02h 42m	+42° 46'	5.8	35'
NGC 869/884	Double Cluster	Per	OC	02h 21m	+57° 08'	4.4	18'
Eta Per		Per	MS	02h 51m	+55° 54'	3.8	29"
M 15		Peg	GC	21h 30m	+12° 10'	6.3	18'
Eps Peg	Enif	Peg	MS	21h 44m	+09° 52'	2.1	143"
Alp Psc	Alrisha	Psc	MS	02h 02m	+02° 46'	3.8	2"
TX Psc		Psc	Var/CS	23h 46m	+03° 29'	4.5-5.3	N/A
55 Psc		Psc	MS	00h 40m	+21° 26'	5.4	6"
Psi1 Psc		Psc	MS	01h 06m	+21° 28'	5.3	30"
Zet Psc		Psc	MS	01h 14m	+07° 35'	5.2	23"
M 33	Triangulum Galaxy	Tri	Gx	01h 34m	+30° 40'	6.4	62'
Alp UMi	Polaris	UMi	MS	02h 51m	+89° 20'	2.0	18"

Planet Visibility Ratings

		Morning Sky							Evening Sky						
		Me	Ve	Ma	Ju	Sa	Ur	Ne	Me	Ve	Ma	Ju	Sa	Ur	Ne
Jan	5th	**	****		*						**			**	**
	15th		****		*						**			**	**
	25th		***		*	**					**			**	*
Feb	5th		***		**	**				*				**	*
	15th		***		**	**				*				*	*
	25th		***		**	**			**	*				*	
Mar	5th		**		**	**				*				*	
	15th		**		***	**				*				*	
	25th	***	**		***	**		*		*				*	
Apr	5th	***	**		***	***		*		*				*	
	15th	***	*		***	***		*		*					
	25th	***	*		****	***		*		*					
May	5th	**	*		****	***		**		*					
	15th		*		****	****	*	**		*					
	25th		*		****	****	*	**		*					
Jun	5th		*		*****	****	*	***	**	*					
	15th		*		****	*		***	***	*	*****				
	25th				****	*		***	****	*	*****				
Jul	5th				*****	**		****	***	*	****				
	15th					**		****		*	****	*****			
	25th					**		****			****	****			
Aug	5th	***				**		****			****	****			
	15th	**				***		*****			***	****			
	25th					***		*****			***	****			
Sep	5th					***		*****			***	****			
	15th					***					**	***			*****
	25th					***					**	***			*****
Oct	5th					***			**		**	***			*****
	15th					****			***	*	**	***			****
	25th			*		****			***	*	**	***			****
Nov	5th			*						*		*	**	****	****
	15th			*						*		*	**	****	****
	25th	***		*						*		*	**	***	***
Dec	5th	***		*						**		*	**	***	***
	15th	**		*						**			**	***	***
	25th			*						**			**	***	**

Solar and Lunar Eclipses

Date	Time (UT)	Type	Visible From
Jan 6th[t]	01:43	Partial Solar	East and north-east Asia and the north Pacific.
Jan 21st	05:11	Total Lunar	Africa, the Atlantic, Europe, the eastern Pacific, North America and South America.
Jul 2nd	19:23	Total Solar	Southern Pacific and southern South America.
Jul 16th	21:30	Partial Lunar	Africa, Antarctica, Asia, the Atlantic, Australia, Europe and South America.
Dec 26th	05:19	Annular	Far eastern Africa, south Asia, the Indian Ocean and the Pacific.

Planetary Highlights

Date	Time (UT)	Elon.	Vis.	Description
Jan 6th	04:44	50° W	AM	Venus is at greatest western elongation from the Sun.
Jan 22nd	05:49	49° W	AM	Venus is 2.4° north of Jupiter.
Feb 13th	20:06	60° E	PM	Mars is 1.1° north of Uranus.
Feb 18th	13:53	44° W	AM	Venus is 1.1° north of Saturn.
Mar 29th	23:51	48° E	PM	Mars is 3.3° south of the Pleiades star cluster. (Taurus.)
Apr 2nd	18:52	23° W	AM	Mercury is 0.4° north of Neptune.
Apr 10th	03:42	30° W	AM	Venus is 0.3° south of Neptune.
Apr 11th	19:29	25° W	AM	Mercury is at greatest western elongation from the Sun.
May 18th	08:09	23° W	AM	Venus is 1.2° south of Uranus.
Jun 10th	16:31	180°	AN	Jupiter is at opposition. (Visible all night.)
Jun 18th	14:41	27° E	PM	Mercury is 0.2° north of Mars.
Jun 23rd	23:03	27° E	PM	Mercury is at greatest eastern elongation from the Sun.
Jul 9th	16:08	180°	AN	Saturn is at opposition. (Visible all night.)
Sep 10th	15:57	180°	AN	Neptune is at opposition. (Visible all night.)
Oct 20th	03:47	23° E	PM	Mercury is at greatest eastern elongation from the Sun.
Oct 30th	08:15	20° E	PM	Mercury is 2.7° south of Venus.
Nov 11th	12:25	0°	N/A	Transit of Mercury across the Sun begins. (Ends 18:04 UT)
Nov 24th	13:57	28° E	PM	Venus is 1.4° south of Jupiter.
Dec 11th	04:34	33° E	PM	Venus is 1.8° south of Saturn.

Major Meteor Showers

Shower Name	Start Date	End Date	Peak	ZHR	Speed	Brightness	Moon
Quadrantids	Dec 28th	Jan 12th	Jan 3rd	120	***	*****	●
Lyrids	Apr 18th	Apr 25th	Apr 22nd	18	***	*****	◑
Eta Aquariids	Apr 24th	May 19th	May 7th	40	*	****	●
Eta Lyrids	May 5th	May 12th	May 10th	3	***	**	●
June Bootids	Jun 23rd	Jun 25th	Jun 24th	Var	*****	*****	◐
Alpha Capricornids	Jul 8th	Aug 10th	Jul 27th	5	*****	****	●
Southern Delta Aquariids	Jul 21st	Aug 23rd	Jul 30th	16	***	*	●
Perseids	Jul 13th	Aug 26th	Aug 12th	100	*	*****	○
Kappa Cygnids	Aug 6th	Aug 31st	Aug 17th	3	*****	**	◑
Aurigids	Aug 29th	Sep 4th	Sep 1st	6	*	****	●
September Epsilon Perseids	Sep 5th	Sep 28th	Sep 9th	5	*	**	○
Draconids	Oct 6th	Oct 10th	Oct 8th	Var	*****	***	○
Southern Taurids	Sep 7th	Nov 19th	Oct 10th	5	****	****	○
Epsilon Geminds	Sep 29th	Nov 2nd	Oct 18th	3	*	**	◐
Orionids	Aug 25th	Nov 19th	Oct 22nd	15	*	****	◑
Andromedids	Oct 26th	Nov 20th	Nov 8th	Var	*****	****	○
Northern Taurids	Oct 25th	Dec 4th	Nov 11th	5	****	****	○
Leonids	Nov 5th	Dec 3rd	Nov 18th	15	*	****	◑
Alpha Monocerotids	Nov 21st	Nov 23rd	Nov 21st	Var	*	****	●
November Orionids	Nov 13th	Dec 7th	Nov 29th	3	***	**	●
Sigma Hydrids	Nov 24th	Dec 21st	Dec 6th	3	*	**	◐
Geminids	Nov 30th	Dec 17th	Dec 13th	120	****	***	◑
December Leonis Minorids	Dec 6th	Jan 18th	Dec 20th	5	*	**	●
Ursids	Dec 17th	Dec 24th	Dec 22nd	10	****	**	●
Coma Berenicids	Dec 24th	Jan 3rd	Dec 31st	5	*	**	●

Key to the Monthly Guide Lunar and Planetary Data Tables

The table below provides a key to the abbreviations and acronyms used in the data tables for the Moon and planets. Please note that the position data (ie, the R.A. and declination) are accurate for 12:00 Universal Time.

+Cr	Waxing Crescent Moon	FQ	First Quarter Moon	LQ	Last Quarter Moon
-Cr	Waning Crescent Moon	+G	Waxing Gibbous Moon	NM	New Moon
FM	Full Moon	-G	Waning Gibbous Moon		
AM	Morning visibility	Gem	Gemini	PM	Evening Sky
Aqr	Aquarius	Ill	Illumination	Psc	Pisces
Ari	Aries	Leo	Leo	R.A.	Right Ascension
Cap	Capricornus	Lib	Libra	Sco	Scorpius
Cet	Cetus	Mag	Magnitude	Sex	Sextans
Con	Constellation	NV	Not Visible	Sgr	Sagittarius
Dec	Declination	Oph	Ophiuchus	Tau	Taurus
Diam	Apparent Diameter	Ori	Orion	Vir	Virgo
Elong	Elongation from Sun				

About the Events

In keeping with the standard used throughout the astronomical community, all of the times are shown in Universal Time and with that in mind, I've included a conversion table below. Also, bear in mind that if an event is not visible at the specific time it occurs (e.g., the Moon appearing close to a planet) it will always be worth observing at the nearest convenient opportunity.

You'll also notice that I don't provide angular separation for conjunctions involving the Moon. Since the Moon is so much closer than the planets, the angular separation will vary depending upon your latitude upon the Earth. With that in mind, I've stuck to stating that the Moon simply appears *close* to a particular planet, star or star cluster in the sky.

	Standard Time	Daylight Savings Time
Greenwich Mean Time	UT-0 hours	UT+1 hours
Eastern Time	UT-5 hours	UT-4 hours
Central Time	UT-6 hours	UT-5 hours
Mountain Time	UT-7 hours	UT-6 hours
Pacific Time	UT-8 hours	UT-7 hours

The Moon

1st   3rd   5th   7th   9th

Date	Con	R.A.	Dec	Mag	Diam	Ill.	Elon.	Phase	Close To
1st	Lib	15h 11m	-12° 43'	-8.3	31'	19%	54° W	-Cr	Venus
2nd	Lib	16h 1m	-16° 14'	-7.6	30'	12%	42° W	NM	Venus, Jupiter, Antares
3rd	Oph	16h 52m	-18° 56'	-6.7	30'	6%	31° W	NM	Jupiter, Antares
4th	Sgr	17h 44m	-20° 43'	-5.7	30'	2%	19° W	NM	Mercury
5th	Sgr	18h 37m	-21° 31'	-4.6	30'	0%	7° W	NM	Mercury, Saturn
6th	Sgr	19h 29m	-21° 18'	-4.4	30'	0%	5° E	NM	Saturn
7th	Cap	20h 19m	-20° 6'	-5.5	30'	2%	17° E	NM	
8th	Cap	21h 9m	-18° 3'	-6.5	29'	5%	28° E	NM	
9th	Cap	21h 57m	-15° 14'	-7.4	29'	10%	39° E	NM	
10th	Aqr	22h 43m	-11° 49'	-8.1	29'	17%	49° E	+Cr	Neptune

Mercury and Venus

Mercury
5th

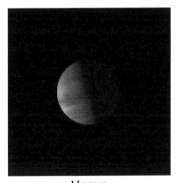

Venus
5th

Mercury

Date	Con.	R.A.	Dec.	Mag.	Diam.	Ill.	Elon.	Vis.	Rat.	Close To
1st	Oph	17h 36m	-23° 16'	-0.4	5"	90%	18° W	AM	**	
3rd	Sgr	17h 49m	-23° 36'	-0.4	5"	91%	17° W	AM	**	
5th	Sgr	18h 2m	-23° 51'	-0.5	5"	92%	16° W	AM	**	Moon
7th	Sgr	18h 15m	-24° 2'	-0.5	5"	94%	14° W	NV	N/A	Saturn
9th	Sgr	18h 28m	-24° 8'	-0.5	5"	95%	13° W	NV	N/A	Saturn

Venus

Date	Con.	R.A.	Dec.	Mag.	Diam.	Ill.	Elon.	Vis.	Rat.	Close To
1st	Lib	15h 29m	-15° 26'	-4.5	26"	48%	49° W	AM	****	Moon
3rd	Lib	15h 37m	-15° 53'	-4.5	26"	49%	49° W	AM	****	
5th	Lib	15h 46m	-16° 21'	-4.4	25"	50%	50° W	AM	****	
7th	Lib	15h 54m	-16° 48'	-4.4	24"	51%	50° W	AM	****	Antares
9th	Lib	16h 2m	-17° 14'	-4.4	24"	52%	50° W	AM	****	Antares

Mars and the Outer Planets

Mars
5th

Jupiter
5th

Saturn
5th

Mars

Date	Con.	R.A.	Dec.	Mag.	Diam.	Ill.	Elon.	Vis.	Rat.	Close To
1st	Psc	0h 1m	0° 9'	0.5	7"	87%	79° E	PM	**	
5th	Psc	0h 11m	0° 60'	0.5	7"	88%	77° E	PM	**	
10th	Psc	0h 23m	2° 26'	0.6	7"	88%	74° E	PM	**	

The Outer Planets

Planet	Date	Con.	R.A.	Dec.	Mag.	Diam.	Elon.	Vis.	Rat.	Close To
Jupiter	5th	Oph	16h 45m	-21° 42'	-1.8	32"	35° W	AM	*	Antares
Saturn	5th	Sgr	18h 51m	-22° 26'	0.5	15"	3° W	NV	N/A	Moon
Uranus	5th	Psc	1h 47m	10° 29'	5.8	4"	101° E	PM	**	
Neptune	5th	Aqr	23h 3m	-7° 7'	7.9	2"	60° E	PM	**	

Highlights

Date	Time (UT)	Event
1st	21:33	The waning crescent Moon is north of Venus. (Morning sky.)
2nd	04:41	Saturn is in conjunction with the Sun. (Not visible.)
	N/A	Good opportunity to see Earthshine on the waning crescent Moon. (Morning sky.)
3rd	00:14	The waning crescent Moon is north of the bright star Antares. (Scorpius, morning sky.)
	06:38	The waning crescent Moon is north of Jupiter. (Morning sky.)
4th	18:59	The waning crescent Moon is north of Mercury. (Morning sky.)
	N/A	The Quadrantid meteor shower is at its maximum. (ZHR: 120)
6th	01:29	New Moon (Not visible.)
	01:43	Partial Solar Eclipse. Visible from east and north-east Asia and the north Pacific.
	04:44	Venus is at greatest western elongation. (Morning sky.)
7th	00:18	Uranus is stationary prior to resuming prograde motion. (Evening sky.)
9th	N/A	Good opportunity to see Earthshine on the waxing crescent Moon. (Evening sky.)
10th	23:36	The waxing crescent Moon is south of Neptune. (Evening sky.)

Planet Locations – January 5th

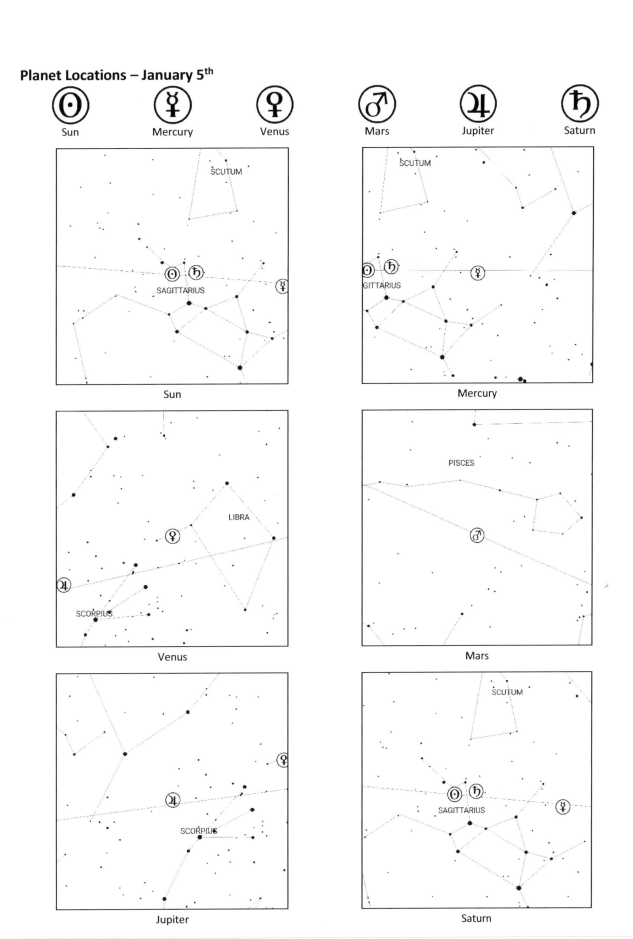

January 11th – 20th

The Moon

 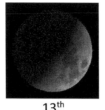  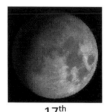 

| 11th | 13th | 15th | 17th | 19th |

Date	Con	R.A.	Dec	Mag	Diam	Ill.	Elon.	Phase	Close To
11th	Aqr	23h 29m	-7° 55'	-8.7	30'	24%	60° E	+Cr	Neptune
12th	Psc	0h 14m	-3° 40'	-9.3	30'	33%	70° E	+Cr	Mars
13th	Cet	0h 59m	0° 47'	-9.8	30'	42%	80° E	FQ	Mars, Uranus
14th	Psc	1h 46m	5° 18'	-10.2	31'	52%	91° E	FQ	Uranus
15th	Cet	2h 34m	9° 42'	-10.6	31'	62%	102° E	FQ	Uranus
16th	Tau	3h 25m	13° 48'	-11.0	31'	72%	113° E	+G	Pleiades
17th	Tau	4h 20m	17° 20'	-11.4	32'	82%	126° E	+G	Pleiades, Hyades, Aldebaran
18th	Tau	5h 19m	19° 58'	-11.7	33'	90%	140° E	+G	Aldebaran
19th	Gem	6h 21m	21° 22'	-12.1	33'	96%	154° E	FM	
20th	Gem	7h 26m	21° 20'	-12.5	33'	99%	169° E	FM	

Mercury and Venus

Mercury
15th

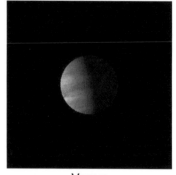

Venus
15th

Mercury

Date	Con.	R.A.	Dec.	Mag.	Diam.	Ill.	Elon.	Vis.	Rat.	Close To
11th	Sgr	18h 42m	-24° 8'	-0.6	5"	95%	12° W	NV	N/A	Saturn
13th	Sgr	18h 55m	-24° 4'	-0.6	5"	96%	11° W	NV	N/A	Saturn
15th	Sgr	19h 9m	-23° 54'	-0.7	5"	97%	10° W	NV	N/A	Saturn
17th	Sgr	19h 23m	-23° 38'	-0.7	5"	98%	8° W	NV	N/A	Saturn
19th	Sgr	19h 37m	-23° 17'	-0.8	5"	98%	7° W	NV	N/A	Saturn

Venus

Date	Con.	R.A.	Dec.	Mag.	Diam.	Ill.	Elon.	Vis.	Rat.	Close To
11th	Sco	16h 11m	-17° 40'	-4.4	23"	53%	50° W	AM	****	Jupiter, Antares
13th	Sco	16h 20m	-18° 5'	-4.4	23"	54%	50° W	AM	****	Jupiter, Antares
15th	Oph	16h 28m	-18° 28'	-4.4	22"	55%	50° W	AM	****	Jupiter, Antares
17th	Oph	16h 37m	-18° 51'	-4.4	22"	56%	50° W	AM	****	Jupiter, Antares
19th	Oph	16h 46m	-19° 12'	-4.3	22"	57%	50° W	AM	***	Jupiter, Antares

Mars and the Outer Planets

Mars
15th

Jupiter
15th

Saturn
15th

Mars

Date	Con.	R.A.	Dec.	Mag.	Diam.	Ill.	Elon.	Vis.	Rat.	Close To
11th	Psc	0h 25m	2° 43'	0.6	7"	88%	74° E	PM	**	
15th	Psc	0h 35m	3° 51'	0.7	7"	88%	72° E	PM	**	
20th	Psc	0h 48m	5° 16'	0.7	7"	89%	70° E	PM	**	

The Outer Planets

Planet	Date	Con.	R.A.	Dec.	Mag.	Diam.	Elon.	Vis.	Rat.	Close To
Jupiter	15th	Oph	16h 53m	-21° 56'	-1.8	33"	43° W	AM	*	Venus, Antares
Saturn	15th	Sgr	18h 56m	-22° 20'	0.5	15"	13° W	NV	N/A	Mercury
Uranus	15th	Psc	1h 47m	10° 30'	5.8	4"	90° E	PM	**	Moon
Neptune	15th	Aqr	23h 4m	-7° 1'	7.9	2"	49° E	PM	**	

Highlights

Date	Time (UT)	Event
11th	23:36	Dwarf planet Pluto is in conjunction with the Sun. (Not visible.)
12th	21:09	The waxing crescent Moon is south of Mars. (Evening sky.)
14th	06:46	First Quarter Moon. (Evening sky.)
	11:15	The first quarter Moon is south of Uranus. (Evening sky.)
15th	14:20	Venus is 7.9° north of the bright star Antares. (Scorpius, morning sky.)
16th	22:06	The waxing gibbous Moon is south of the Pleaides star cluster. (Taurus, evening sky.)
17th	17:23	The waxing gibbous Moon is north of the bright star Aldebaran. (Taurus, evening sky.)

Planet Locations – January 15th

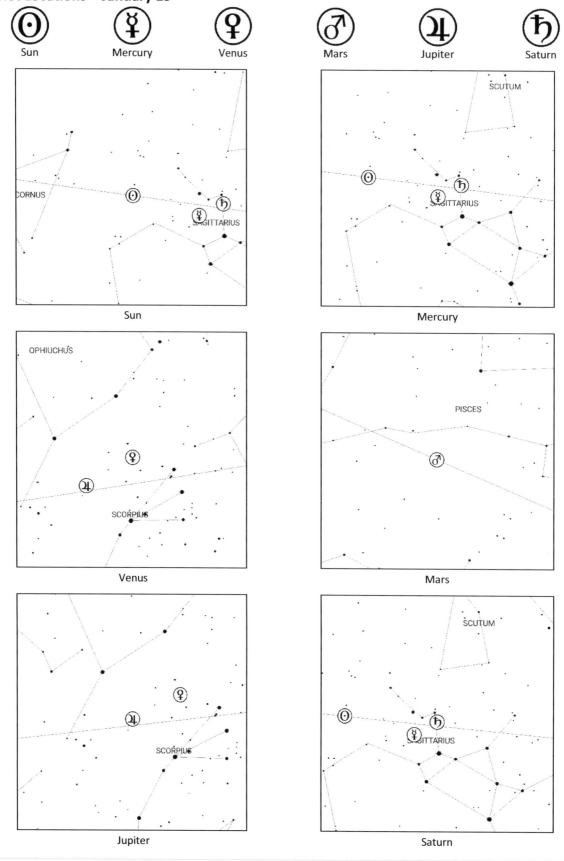

Sun

Mercury

Venus

Mars

Jupiter

Saturn

The Moon

| 21st | 23rd | 25th | 27th | 29th | 31st |

Date	Con	R.A.	Dec	Mag	Diam	Ill.	Elon.	Phase	Close To
21st	Cnc	8h 30m	19° 44'	-12.6	33'	100%	176° W	FM	Praesepe
22nd	Leo	9h 33m	16° 43'	-12.3	33'	97%	161° W	FM	Regulus
23rd	Leo	10h 33m	12° 36'	-11.9	33'	92%	147° W	-G	Regulus
24th	Leo	11h 30m	7° 46'	-11.5	33'	85%	134° W	-G	
25th	Vir	12h 25m	2° 37'	-11.1	32'	75%	121° W	-G	
26th	Vir	13h 17m	-2° 32'	-10.7	32'	65%	109° W	LQ	Spica
27th	Vir	14h 8m	-7° 24'	-10.3	31'	54%	97° W	LQ	Spica
28th	Lib	14h 59m	-11° 46'	-9.8	31'	43%	86° W	LQ	
29th	Lib	15h 49m	-15° 28'	-9.3	31'	33%	74° W	-Cr	Antares
30th	Oph	16h 40m	-18° 22'	-8.7	30'	24%	63° W	-Cr	Jupiter, Antares
31st	Oph	17h 32m	-20° 23'	-8.1	30'	16%	51° W	-Cr	Venus, Jupiter

Mercury and Venus

Mercury
25th

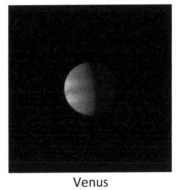

Venus
25th

Mercury

Date	Con.	R.A.	Dec.	Mag.	Diam.	Ill.	Elon.	Vis.	Rat.	Close To
21st	Sgr	19h 51m	-22° 50'	-0.9	5"	99%	6° W	NV	N/A	
23rd	Sgr	20h 5m	-22° 17'	-1.0	5"	99%	4° W	NV	N/A	
25th	Cap	20h 19m	-21° 38'	-1.1	5"	99%	3° W	NV	N/A	
27th	Cap	20h 33m	-20° 54'	-1.2	5"	100%	1° W	NV	N/A	
29th	Cap	20h 47m	-20° 3'	-1.3	5"	100%	0° E	NV	N/A	
31st	Cap	21h 1m	-19° 6'	-1.4	5"	100%	2° E	NV	N/A	

Venus

Date	Con.	R.A.	Dec.	Mag.	Diam.	Ill.	Elon.	Vis.	Rat.	Close To
21st	Oph	16h 56m	-19° 32'	-4.3	21"	58%	49° W	AM	***	Jupiter, Antares
23rd	Oph	17h 5m	-19° 50'	-4.3	21"	59%	49° W	AM	***	Jupiter, Antares
25th	Oph	17h 14m	-20° 7'	-4.3	20"	59%	49° W	AM	***	Jupiter
27th	Oph	17h 24m	-20° 22'	-4.3	20"	60%	49° W	AM	***	Jupiter
29th	Oph	17h 33m	-20° 35'	-4.3	20"	61%	48° W	AM	***	Jupiter
31st	Oph	17h 43m	-20° 46'	-4.3	19"	62%	48° W	AM	***	Moon, Jupiter

Mars and the Outer Planets

Mars
25th

Jupiter
25th

Saturn
25th

Mars

Date	Con.	R.A.	Dec.	Mag.	Diam.	Ill.	Elon.	Vis.	Rat.	Close To
21st	Psc	0h 50m	5° 33'	0.7	7"	89%	69° E	PM	**	
25th	Psc	1h 0m	6° 39'	0.8	6"	89%	68° E	PM	**	
31st	Psc	1h 15m	8° 17'	0.9	6"	89%	65° E	PM	*	

The Outer Planets

Planet	Date	Con.	R.A.	Dec.	Mag.	Diam.	Elon.	Vis.	Rat.	Close To
Jupiter	25th	Oph	17h 2m	-22° 9'	-1.9	33"	52° W	AM	*	Venus, Antares
Saturn	25th	Sgr	19h 1m	-22° 13'	0.5	15"	22° W	AM	**	
Uranus	25th	Psc	1h 47m	10° 33'	5.8	4"	79° E	PM	**	
Neptune	25th	Aqr	23h 5m	-6° 54'	7.9	2"	39° E	PM	*	

Highlights

Date	Time (UT)	Event
21st	05:11	Total lunar eclipse. Visible from Africa, the Atlantic, Europe, the eastern Pacific, North America and South America.
	05:17	Full Moon. (Visible all night.)
	14:51	The full Moon is south of the Praesepe star cluster. (Cancer, visible all night.)
22nd	05:49	Venus is 2.4° north of Jupiter. (Morning sky.)
23rd	01:35	The waning gibbous Moon is north of the bright star Regulus. (Leo, morning sky.)
26th	15:54	The nearly last quarter Moon is north of the bright star Spica. (Virgo, morning sky.)
27th	21:11	Last Quarter Moon. (Morning sky.)
30th	02:36	Mercury is at superior conjunction with the Sun. (Not visible.)
	05:59	The waning crescent Moon is north of the bright star Antares. (Scorpius, morning sky.)
	23:03	The waning crescent Moon is north of Jupiter. (Morning sky.)
31st	18:28	The waning crescent Moon is south of Venus. (Morning sky.)

Planet Locations – January 25th

☉ Sun   ☿ Mercury   ♀ Venus   ♂ Mars   ♃ Jupiter   ♄ Saturn

Sun

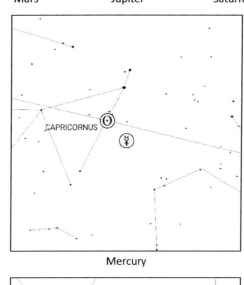

Mercury

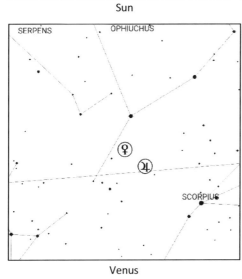

Venus

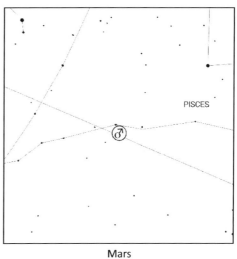

Mars

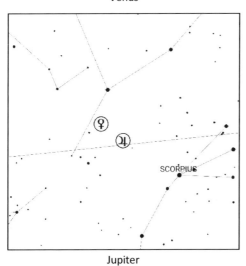

Jupiter

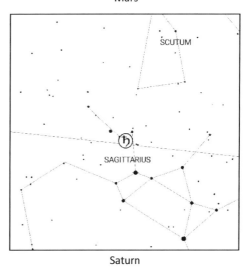

Saturn

February 1st to 10th

The Moon

 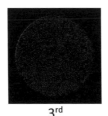   

1st   3rd   5th   7th   9th

Date	Con	R.A.	Dec	Mag	Diam	Ill.	Elon.	Phase	Close To
1st	Sgr	18h 24m	-21° 24'	-7.3	30'	10%	39° W	NM	Venus, Saturn
2nd	Sgr	19h 15m	-21° 26'	-6.5	30'	5%	27° W	NM	Saturn
3rd	Sgr	20h 6m	-20° 30'	-5.5	29'	2%	15° W	NM	
4th	Cap	20h 56m	-18° 39'	-4.3	29'	0%	4° W	NM	Mercury
5th	Cap	21h 44m	-16° 1'	-4.7	29'	0%	7° E	NM	Mercury
6th	Aqr	22h 31m	-12° 44'	-5.7	29'	2%	18° E	NM	Neptune
7th	Aqr	23h 17m	-8° 56'	-6.7	29'	6%	29° E	NM	Neptune
8th	Psc	0h 2m	-4° 46'	-7.5	30'	11%	39° E	NM	
9th	Cet	0h 47m	0° 23'	-8.2	30'	18%	49° E	+Cr	
10th	Psc	1h 32m	4° 5'	-8.9	30'	26%	59° E	+Cr	Mars, Uranus

Mercury and Venus

Mercury
5th

Venus
5th

Mercury

Date	Con.	R.A.	Dec.	Mag.	Diam.	Ill.	Elon.	Vis.	Rat.	Close To
1st	Cap	21h 8m	-18° 36'	-1.4	5"	100%	2° E	NV	N/A	
3rd	Cap	21h 22m	-17° 30'	-1.4	5"	99%	4° E	NV	N/A	
5th	Cap	21h 36m	-16° 18'	-1.3	5"	99%	5° E	NV	N/A	Moon
7th	Cap	21h 50m	-15° 1'	-1.3	5"	98%	7° E	NV	N/A	
9th	Aqr	22h 4m	-13° 38'	-1.3	5"	96%	8° E	NV	N/A	

Venus

Date	Con.	R.A.	Dec.	Mag.	Diam.	Ill.	Elon.	Vis.	Rat.	Close To
1st	Sgr	17h 48m	-20° 51'	-4.3	19"	62%	48° W	AM	***	Moon
3rd	Sgr	17h 57m	-20° 60'	-4.2	19"	63%	47° W	AM	***	
5th	Sgr	18h 7m	-21° 6'	-4.2	19"	64%	47° W	AM	***	
7th	Sgr	18h 17m	-21° 10'	-4.2	18"	65%	46° W	AM	***	
9th	Sgr	18h 27m	-21° 12'	-4.2	18"	65%	46° W	AM	***	

Mars and the Outer Planets

Mars
5th

Jupiter
5th

Saturn
5th

Mars

Date	Con.	R.A.	Dec.	Mag.	Diam.	Ill.	Elon.	Vis.	Rat.	Close To
1st	Psc	1h 18m	8° 34'	0.9	6"	89%	65° E	PM	*	
5th	Psc	1h 28m	9° 37'	0.9	6"	90%	63° E	PM	*	
10th	Psc	1h 40m	10° 55'	1.0	6"	90%	61° E	PM	*	Moon, Uranus

The Outer Planets

Planet	Date	Con.	R.A.	Dec.	Mag.	Diam.	Elon.	Vis.	Rat.	Close To
Jupiter	5th	Oph	17h 10m	-22° 19'	-1.9	34"	61° W	AM	**	
Saturn	5th	Sgr	19h 6m	-22° 6'	0.6	15"	32° W	AM	**	
Uranus	5th	Psc	1h 48m	10° 38'	5.8	4"	68° E	PM	**	Mars
Neptune	5th	Aqr	23h 6m	-6° 46'	8.0	2"	28° E	PM	*	

Highlights

Date	Time (UT)	Event
1st	N/A	Good opportunity to see Earthshine on the waning crescent Moon. (Morning sky.)
2nd	05:57	The waning crescent Moon is south of Saturn. (Morning sky.)
4th	21:04	New Moon. (Not visible.)
7th	05:15	The waxing crescent Moon appears south of Neptune. (Evening sky.)
8th	N/A	Good opportunity to see Earthshine on the waxing crescent Moon. (Evening sky.)
10th	16:29	The waxing crescent Moon is south of Mars. (Evening sky.)
	21:46	The waxing crescent Moon Is south of Uranus. (Evening sky.)

Planet Locations – February 5th

 Sun  Mercury   Venus   Mars Jupiter Saturn

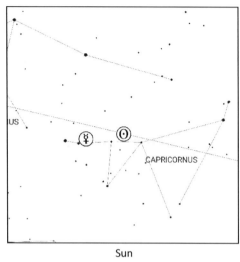
Sun

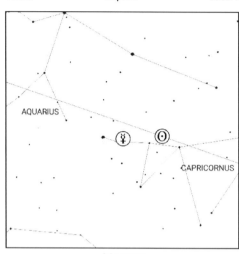
Mercury

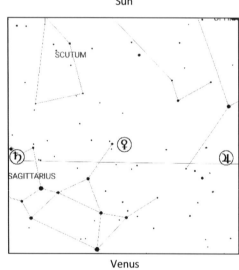
Venus

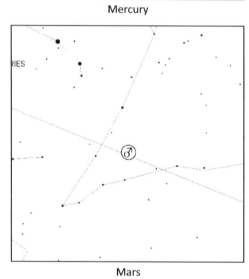
Mars

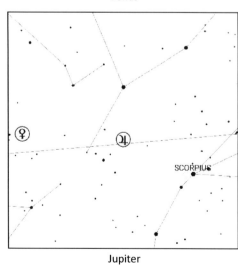
Jupiter

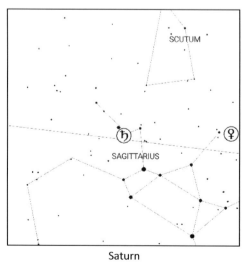
Saturn

The Moon

| 11th | 13th | 15th | 17th | 19th |

Date	Con	R.A.	Dec	Mag	Diam	Ill.	Elon.	Phase	Close To
11th	Cet	2h 19m	8° 28'	-9.4	30'	35%	70° E	+Cr	Mars, Uranus
12th	Ari	3h 8m	12° 35'	-9.9	31'	45%	81° E	FQ	Pleiades
13th	Tau	4h 0m	16° 13'	-10.4	31'	56%	93° E	FQ	Pleiades, Hyades, Aldebaran
14th	Tau	4h 55m	19° 7'	-10.8	32'	67%	106° E	+G	Hyades, Aldebaran
15th	Ori	5h 54m	20° 60'	-11.2	32'	77%	120° E	+G	
16th	Gem	6h 56m	21° 35'	-11.6	33'	86%	134° E	+G	
17th	Gem	8h 0m	20° 42'	-11.9	33'	93%	149° E	+G	Praesepe
18th	Cnc	9h 3m	18° 20'	-12.3	33'	98%	164° E	FM	Praesepe
19th	Leo	10h 5m	14° 40'	-12.6	33'	100%	179° E	FM	Regulus
20th	Leo	11h 5m	10° 1'	-12.4	33'	99%	167° W	FM	

Mercury and Venus

Mercury
15th

Venus
15th

Mercury

Date	Con.	R.A.	Dec.	Mag.	Diam.	Ill.	Elon.	Vis.	Rat.	Close To
11th	Aqr	22h 17m	-12° 11'	-1.2	5"	94%	10° E	NV	N/A	
13th	Aqr	22h 31m	-10° 38'	-1.2	5"	92%	11° E	NV	N/A	
15th	Aqr	22h 44m	-9° 3'	-1.1	6"	88%	12° E	NV	N/A	
17th	Aqr	22h 56m	-7° 25'	-1.1	6"	84%	13° E	NV	N/A	Neptune
19th	Aqr	23h 8m	-5° 46'	-1.0	6"	78%	15° E	NV	N/A	Neptune

Venus

Date	Con.	R.A.	Dec.	Mag.	Diam.	Ill.	Elon.	Vis.	Rat.	Close To
11th	Sgr	18h 37m	-21° 11'	-4.2	18"	66%	46° W	AM	***	Saturn
13th	Sgr	18h 47m	-21° 9'	-4.2	17"	67%	45° W	AM	***	Saturn
15th	Sgr	18h 57m	-21° 4'	-4.2	17"	68%	44° W	AM	***	Saturn
17th	Sgr	19h 7m	-20° 56'	-4.2	17"	68%	44° W	AM	***	Saturn
19th	Sgr	19h 17m	-20° 47'	-4.1	17"	69%	43° W	AM	***	Saturn

Mars and the Outer Planets

Mars
15th

Jupiter
15th

Saturn
15th

Mars

Date	Con.	R.A.	Dec.	Mag.	Diam.	Ill.	Elon.	Vis.	Rat.	Close To
11th	Psc	1h 43m	11° 11'	1.0	6"	90%	61° E	PM	*	Moon, Uranus
15th	Ari	1h 53m	12° 11'	1.0	6"	90%	60° E	PM	*	Uranus
20th	Ari	2h 6m	13° 24'	1.1	6"	91%	58° E	PM	*	Uranus

The Outer Planets

Planet	Date	Con.	R.A.	Dec.	Mag.	Diam.	Elon.	Vis.	Rat.	Close To
Jupiter	15th	Oph	17h 17m	-22° 27'	-2.0	35"	69° W	AM	**	
Saturn	15th	Sgr	19h 11m	-21° 59'	0.6	15"	41° W	AM	**	Venus
Uranus	15th	Ari	1h 49m	10° 44'	5.8	3"	59° E	PM	*	Mars
Neptune	15th	Aqr	23h 7m	-6° 38'	8.0	2"	18° E	PM	*	Mercury

Highlights

Date	Time (UT)	Event
12th	22:27	First Quarter Moon. (Evening sky.)
13th	06:00	The just-past first quarter Moon is south of the Pleiades star cluster. (Taurus, evening sky.)
	20:06	Mars is 1.1° north of Uranus. (Evening sky.)
14th	04:20	The waxing gibbous Moon is north of the bright star Aldebaran. (Taurus, evening sky.)
18th	04:05	The nearly full Moon is south of the Praesepe star cluster. (Cancer, evening sky.)
	13:53	Venus is 1.1° north of Saturn. (Morning sky.)
19th	12:51	The almost full Moon is north of the bright star Regulus. (Leo, evening sky.)
	15:54	Full Moon. (Evening sky.)

Planet Locations – February 15th

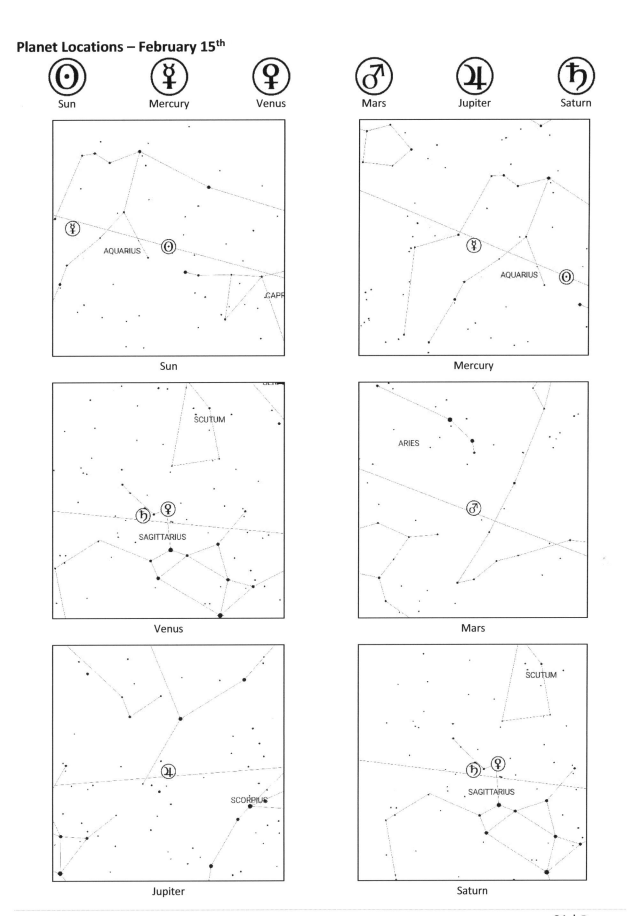

February 21st – 28th

The Moon

| 21st | 23rd | 25th | 27th |

Date	Con	R.A.	Dec	Mag	Diam	Ill.	Elon.	Phase		Close To
21st	Vir	12h 2m	4° 47'	-12.0	33'	95%	154° W	-G		
22nd	Vir	12h 56m	0° 36'	-11.7	33'	88%	141° W	-G		Spica
23rd	Vir	13h 50m	-5° 48'	-11.3	32'	80%	129° W	-G		Spica
24th	Lib	14h 42m	-10° 32'	-10.9	32'	70%	117° W	-G		
25th	Lib	15h 34m	-14° 35'	-10.5	31'	60%	105° W	LQ		
26th	Oph	16h 26m	-17° 48'	-10.1	31'	50%	93° W	LQ		Antares
27th	Oph	17h 19m	-20° 4'	-9.7	30'	40%	80° W	LQ		Jupiter
28th	Sgr	18h 11m	-21° 21'	-9.1	30'	30%	68° W	-Cr		Jupiter

Mercury and Venus

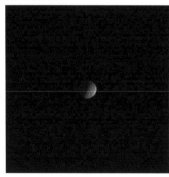 

Mercury
24th

Venus
24th

Mercury

Date	Con.	R.A.	Dec.	Mag.	Diam.	Ill.	Elon.	Vis.	Rat.		Close To
21st	Aqr	23h 19m	-4° 9'	-0.9	6"	72%	15° E	PM	**		Neptune
23rd	Psc	23h 29m	-2° 35'	-0.8	7"	64%	16° E	PM	**		
25th	Psc	23h 38m	-1° 9'	-0.6	7"	55%	16° E	PM	**		
27th	Psc	23h 45m	0° 8'	-0.3	7"	46%	16° E	PM	**		

Venus

Date	Con.	R.A.	Dec.	Mag.	Diam.	Ill.	Elon.	Vis.	Rat.	Close To
21st	Sgr	19h 27m	-20° 35'	-4.1	16"	70%	43° W	AM	***	Saturn
23rd	Sgr	19h 37m	-20° 21'	-4.1	16"	70%	42° W	AM	***	Saturn
25th	Sgr	19h 47m	-20° 4'	-4.1	16"	71%	42° W	AM	***	Saturn
27th	Sgr	19h 57m	-19° 46'	-4.1	16"	72%	41° W	AM	**	

Mars and the Outer Planets

Mars	Jupiter	Saturn
24th	24th	24th

Mars

Date	Con.	R.A.	Dec.	Mag.	Diam.	Ill.	Elon.	Vis.	Rat.	Close To
21st	Ari	2h 9m	13° 39'	1.1	6"	91%	58° E	PM	*	Uranus
24th	Ari	2h 16m	14° 21'	1.1	5"	91%	57° E	PM	*	
28th	Ari	2h 27m	15° 16'	1.2	5"	91%	56° E	PM	*	

The Outer Planets

Planet	Date	Con.	R.A.	Dec.	Mag.	Diam.	Elon.	Vis.	Rat.	Close To
Jupiter	24th	Oph	17h 22m	-22° 32'	-2.0	36"	77° W	AM	**	
Saturn	24th	Sgr	19h 14m	-21° 53'	0.6	16"	49° W	AM	**	Venus
Uranus	24th	Ari	1h 50m	10° 51'	5.8	3"	50° E	PM	*	Mars
Neptune	24th	Aqr	23h 9m	-6° 30'	8.0	2"	10° E	NV	N/A	Mercury

Highlights

Date	Time (UT)	Event
22nd	23:35	The waning gibbous Moon is north of the bright star Spica. (Virgo, morning sky.)
26th	11:28	Last Quarter Moon. (Morning sky.)
	14:11	The last quarter Moon is north of the bright star Antares. (Scorpius, evening sky.)
27th	01:14	Mercury is at greatest eastern elongation. (Evening sky.)
	15:18	The just-past last quarter Moon is north of Jupiter. (Morning sky.)

Planet Locations – February 25th

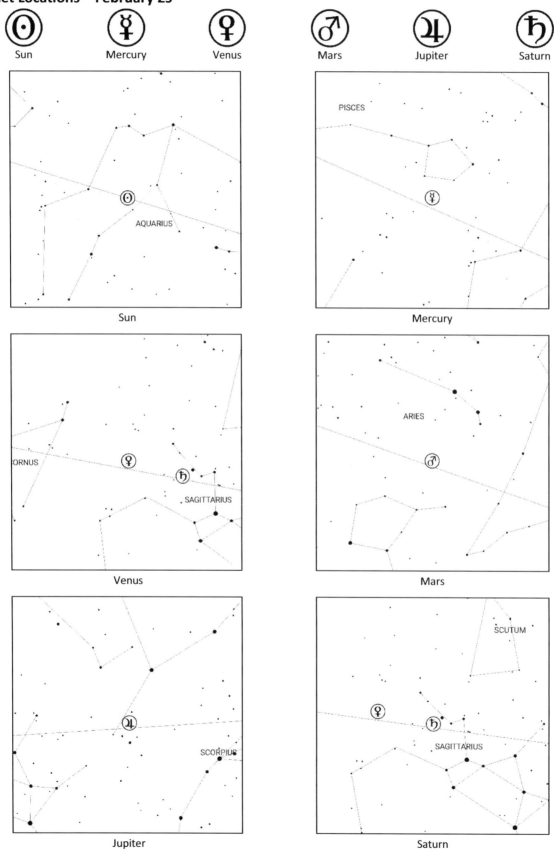

The Moon

	1st		3rd		5th		7th		9th

Date	Con	R.A.	Dec	Mag	Diam	Ill.	Elon.	Phase	Close To
1st	Sgr	19h 3m	-21° 37'	-8.6	30'	22%	56° W	-Cr	Saturn
2nd	Sgr	19h 54m	-20° 53'	-7.9	30'	15%	45° W	-Cr	Venus, Saturn
3rd	Cap	20h 44m	-19° 15'	-7.1	29'	9%	33° W	NM	Venus
4th	Cap	21h 32m	-16° 47'	-6.3	29'	4%	22° W	NM	
5th	Aqr	22h 19m	-13° 37'	-5.3	29'	1%	11° W	NM	
6th	Aqr	23h 6m	-9° 54'	-4.3	30'	0%	0° W	NM	Mercury, Neptune
7th	Aqr	23h 51m	-5° 45'	-5.0	30'	1%	10° E	NM	Mercury, Neptune
8th	Cet	0h 36m	-1° 21'	-6.0	30'	3%	21° E	NM	Mercury
9th	Psc	1h 21m	3° 9'	-7.0	30'	8%	31° E	NM	Uranus
10th	Cet	2h 8m	7° 36'	-7.8	30'	14%	42° E	+Cr	Mars, Uranus

Mercury and Venus

Mercury
5th

Venus
5th

Mercury

Date	Con.	R.A.	Dec.	Mag.	Diam.	Ill.	Elon.	Vis.	Rat.	Close To
1st	Psc	23h 50m	1° 13'	0.0	8"	37%	15° E	PM	**	
3rd	Psc	23h 53m	2° 2'	0.5	8"	28%	14° E	NV	N/A	
5th	Psc	23h 53m	2° 34'	1.1	9"	20%	13° E	NV	N/A	
7th	Psc	23h 52m	2° 47'	1.8	9"	13%	10° E	NV	N/A	Moon
9th	Psc	23h 49m	2° 40'	2.6	10"	7%	8° E	NV	N/A	

Venus

Date	Con.	R.A.	Dec.	Mag.	Diam.	Ill.	Elon.	Vis.	Rat.	Close To
1st	Sgr	20h 7m	-19° 25'	-4.1	16"	72%	40° W	AM	**	
3rd	Cap	20h 16m	-19° 1'	-4.1	15"	73%	40° W	AM	**	Moon
5th	Cap	20h 26m	-18° 36'	-4.1	15"	74%	39° W	AM	**	
7th	Cap	20h 36m	-18° 9'	-4.1	15"	74%	39° W	AM	**	
9th	Cap	20h 46m	-17° 39'	-4.1	15"	75%	38° W	AM	**	

Mars and the Outer Planets

Mars
5th

Jupiter
5th

Saturn
5th

Mars

Date	Con.	R.A.	Dec.	Mag.	Diam.	Ill.	Elon.	Vis.	Rat.	Close To
1st	Ari	2h 30m	15° 29'	1.2	5"	91%	55° E	PM	*	
5th	Ari	2h 40m	16° 21'	1.2	5"	92%	54° E	PM	*	
10th	Ari	2h 53m	17° 24'	1.3	5"	92%	53° E	PM	*	Moon

The Outer Planets

Planet	Date	Con.	R.A.	Dec.	Mag.	Diam.	Elon.	Vis.	Rat.	Close To
Jupiter	5th	Oph	17h 26m	-22° 36'	-2.1	37"	84° W	AM	**	
Saturn	5th	Sgr	19h 18m	-21° 47'	0.6	16"	56° W	AM	**	
Uranus	5th	Ari	1h 52m	10° 59'	5.9	3"	42° E	PM	*	
Neptune	5th	Aqr	23h 10m	-6° 22'	8.0	2"	2° E	NV	N/A	

Highlights

Date	Time (UT)	Event
1st	19:01	The waning crescent Moon is south of Saturn. (Morning sky.)
2nd	21:30	The waning crescent Moon is south of Venus. (Morning sky.)
5th	05:19	Mercury is stationary prior to beginning retrograde motion. (Not visible.)
6th	16:05	New Moon. (Not visible.)
7th	11:08	Neptune is in conjunction with the Sun. (Not visible.)
9th	N/A	Good opportunity to see Earthshine on the waxing crescent Moon. (Evening sky.)
10th	03:49	The waxing crescent Moon is south of Uranus. (Evening sky.)

Planet Locations – March 5th

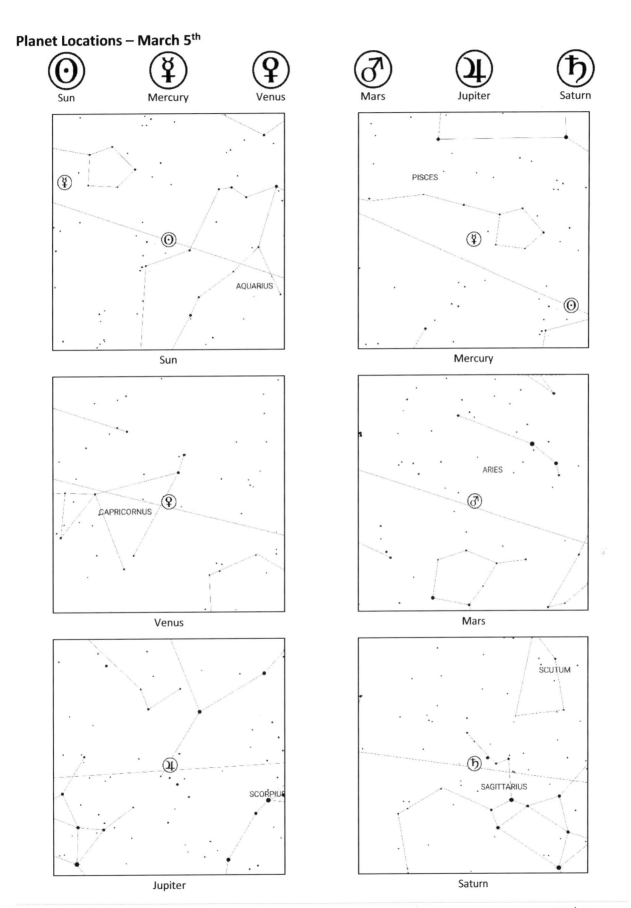

March 11ᵗʰ – 20ᵗʰ

The Moon

| 11ᵗʰ | 13ᵗʰ | 15ᵗʰ | 17ᵗʰ | 19ᵗʰ |

Date	Con	R.A.	Dec	Mag	Diam	Ill.	Elon.	Phase	Close To
11th	Ari	2h 56m	11° 47'	-8.5	31'	21%	53° E	+Cr	Mars
12th	Tau	3h 46m	15° 32'	-9.1	31'	30%	64° E	+Cr	Mars, Pleiades, Hyades
13th	Tau	4h 39m	18° 35'	-9.7	31'	40%	77° E	FQ	Hyades, Aldebaran
14th	Tau	5h 36m	20° 44'	-10.2	32'	51%	90° E	FQ	
15th	Gem	6h 35m	21° 43'	-10.6	32'	62%	104° E	FQ	
16th	Gem	7h 35m	21° 22'	-11.0	32'	72%	118° E	+G	
17th	Cnc	8h 37m	19° 37'	-11.4	33'	82%	132° E	+G	Praesepe
18th	Leo	9h 38m	16° 30'	-11.8	33'	90%	147° E	+G	Regulus
19th	Leo	10h 37m	12° 17'	-12.1	33'	96%	161° E	FM	Regulus
20th	Leo	11h 35m	7° 15'	-12.5	33'	99%	174° E	FM	

Mercury and Venus

Mercury
15ᵗʰ

Venus
15ᵗʰ

Mercury

Date	Con.	R.A.	Dec.	Mag.	Diam.	Ill.	Elon.	Vis.	Rat.	Close To
11th	Psc	23h 44m	2° 16'	3.5	10"	3%	5° E	NV	N/A	
13th	Psc	23h 38m	1° 35'	4.5	11"	1%	1° E	NV	N/A	
15th	Psc	23h 31m	0° 42'	4.9	11"	1%	2° W	NV	N/A	Neptune
17th	Psc	23h 25m	0° 18'	4.3	11"	2%	6° W	NV	N/A	Neptune
19th	Psc	23h 19m	-1° 21'	3.6	11"	4%	9° W	NV	N/A	Neptune

Venus

Date	Con.	R.A.	Dec.	Mag.	Diam.	Ill.	Elon.	Vis.	Rat.	Close To
11th	Cap	20h 56m	-17° 7'	-4.0	15"	75%	37° W	AM	**	
13th	Cap	21h 5m	-16° 34'	-4.0	14"	76%	37° W	AM	**	
15th	Cap	21h 15m	-15° 58'	-4.0	14"	77%	36° W	AM	**	
17th	Cap	21h 25m	-15° 21'	-4.0	14"	77%	36° W	AM	**	
19th	Cap	21h 34m	-14° 42'	-4.0	14"	78%	35° W	AM	**	

Mars and the Outer Planets

Mars	Jupiter	Saturn
15th	15th	15th

Mars

Date	Con.	R.A.	Dec.	Mag.	Diam.	Ill.	Elon.	Vis.	Rat.	Close To
11th	Ari	2h 56m	17° 36'	1.3	5"	92%	53° E	PM	*	Moon
15th	Ari	3h 7m	18° 22'	1.3	5"	93%	52° E	PM	*	Pleiades
20th	Ari	3h 20m	19° 18'	1.4	5"	93%	51° E	PM	*	Pleiades

The Outer Planets

Planet	Date	Con.	R.A.	Dec.	Mag.	Diam.	Elon.	Vis.	Rat.	Close To
Jupiter	15th	Oph	17h 30m	-22° 38'	-2.1	38"	92° W	AM	***	
Saturn	15th	Sgr	19h 21m	-21° 41'	0.6	16"	65° W	AM	**	
Uranus	15th	Ari	1h 53m	11° 9'	5.9	3"	33° E	PM	*	
Neptune	15th	Aqr	23h 11m	-6° 13'	8.0	2"	7° W	NV	N/A	Mercury

Highlights

Date	Time (UT)	Event
11th	08:30	Asteroid Vesta is in conjunction with the Sun. (Not visible.)
	10:51	The waxing crescent Moon is south of Mars. (Evening sky.)
12th	10:55	The waxing crescent Moon is south of the Pleaides star cluster. (Taurus, evening sky.)
13th	09:13	The nearly first quarter Moon is north of the bright star Aldebaran. (Taurus, evening sky.)
14th	10:28	First Quarter Moon. (Evening sky.)
15th	01:42	Mercury is at inferior conjunction with the Sun. (Not visible.)
17th	12:05	The waxing gibbous Moon is south of the Praesepe star cluster. (Cancer, evening sky.)
19th	00:31	The waxing gibbous Moon is north of the bright star Regulus. (Leo, evening sky.)
20th	21:59	Spring equinox.

Planet Locations – March 15ᵗʰ

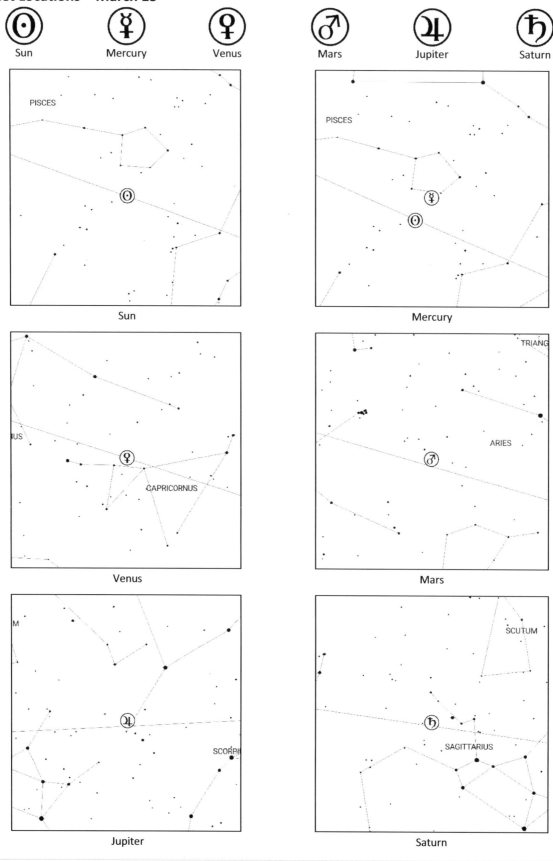

The Moon

| 21ˢᵗ | 23ʳᵈ | 25ᵗʰ | 27ᵗʰ | 29ᵗʰ | 31ˢᵗ |

Date	Con	R.A.	Dec	Mag	Diam	Ill.	Elon.	Phase	Close To
21st	Vir	12h 31m	1° 48'	-12.5	33'	100%	173° W	FM	
22nd	Vir	13h 26m	-3° 39'	-12.2	33'	97%	160° W	FM	Spica
23rd	Vir	14h 20m	-8° 47'	-11.9	32'	92%	147° W	-G	
24th	Lib	15h 14m	-13° 18'	-11.5	32'	84%	135° W	-G	
25th	Sco	16h 8m	-16° 58'	-11.2	31'	76%	122° W	-G	Antares
26th	Oph	17h 1m	-19° 40'	-10.8	31'	66%	110° W	-G	Jupiter, Antares
27th	Sgr	17h 55m	-21° 18'	-10.4	30'	57%	97° W	LQ	Jupiter
28th	Sgr	18h 48m	-21° 52'	-10.0	30'	47%	85° W	LQ	Saturn
29th	Sgr	19h 40m	-21° 24'	-9.5	30'	37%	73° W	-Cr	Saturn
30th	Cap	20h 30m	-19° 58'	-9.0	30'	28%	61° W	-Cr	
31st	Cap	21h 19m	-17° 41'	-8.4	29'	20%	50° W	-Cr	

Mercury and Venus

Mercury
25ᵗʰ

Venus
25ᵗʰ

Mercury

Date	Con.	R.A.	Dec.	Mag.	Diam.	Ill.	Elon.	Vis.	Rat.	Close To
21st	Psc	23h 14m	-2° 21'	2.9	11"	7%	12° W	NV	N/A	Neptune
23rd	Aqr	23h 10m	-3° 17'	2.3	11"	11%	15° W	NV	N/A	Neptune
25th	Aqr	23h 8m	-4° 4'	1.8	10"	15%	17° W	AM	***	Neptune
27th	Aqr	23h 7m	-4° 41'	1.5	10"	19%	19° W	AM	***	Neptune
29th	Aqr	23h 8m	-5° 9'	1.2	10"	24%	21° W	AM	***	Neptune
31st	Aqr	23h 10m	-5° 27'	1.0	9"	28%	22° W	AM	***	Venus, Neptune

Venus

Date	Con.	R.A.	Dec.	Mag.	Diam.	Ill.	Elon.	Vis.	Rat.	Close To
21st	Cap	21h 44m	-14° 2'	-4.0	14"	78%	34° W	AM	**	
23rd	Cap	21h 53m	-13° 19'	-4.0	14"	79%	34° W	AM	**	
25th	Aqr	22h 3m	-12° 36'	-4.0	14"	79%	33° W	AM	**	
27th	Aqr	22h 12m	-11° 50'	-4.0	13"	80%	33° W	AM	**	
29th	Aqr	22h 21m	-11° 4'	-4.0	13"	80%	32° W	AM	**	
31st	Aqr	22h 31m	-10° 16'	-4.0	13"	81%	32° W	AM	**	Mercury

Mars and the Outer Planets

Mars
25th

Jupiter
25th

Saturn
25th

Mars

Date	Con.	R.A.	Dec.	Mag.	Diam.	Ill.	Elon.	Vis.	Rat.	Close To
21st	Ari	3h 23m	19° 28'	1.4	5"	93%	50° E	PM	*	Pleiades
25th	Tau	3h 34m	20° 9'	1.4	5"	93%	49° E	PM	*	Pleiades
31st	Tau	3h 51m	21° 5'	1.4	5"	94%	48° E	PM	*	Pleiades, Hyades

The Outer Planets

Planet	Date	Con.	R.A.	Dec.	Mag.	Diam.	Elon.	Vis.	Rat.	Close To
Jupiter	25th	Oph	17h 33m	-22° 40'	-2.2	39"	101° W	AM	***	
Saturn	25th	Sgr	19h 24m	-21° 36'	0.6	16"	73° W	AM	**	
Uranus	25th	Ari	1h 55m	11° 20'	5.9	3"	25° E	PM	*	
Neptune	25th	Aqr	23h 13m	-6° 5'	8.0	2"	16° W	AM	*	Mercury

Highlights

Date	Time (UT)	Event
21st	01:43	Full Moon. (Visible all night.)
22nd	11:52	The just-past full Moon is north of the bright star Spica. (Virgo, morning sky.)
25th	20:24	The waning gibbous Moon is north of the bright star Antares. (Scorpius, morning sky.)
27th	01:23	The nearly last quarter Moon is north of Jupiter. (Morning sky.)
	11:33	Mercury is stationary prior to resuming prograde motion. (Morning sky.)
28th	04:10	Last Quarter Moon. (Morning sky.)
29th	04:08	The just-past last quarter Moon is south of Saturn. (Morning sky.)
	12:29	The just-past last quarter Moon is south of dwarf planet Pluto. (Morning sky.)
	23:51	Mars is 3.3° south of the Pleiades star cluster. (Taurus, evening sky.)

Planet Locations – March 25th

 Sun  Mercury Venus Mars Jupiter Saturn

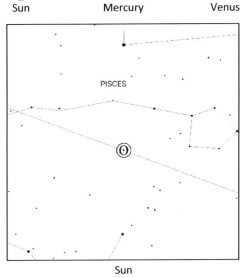

Sun

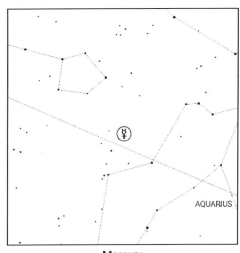

Mercury

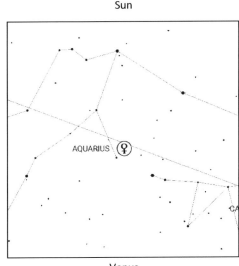

Venus

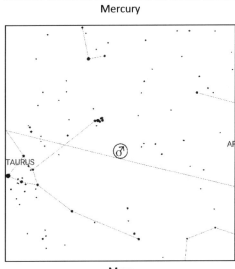

Mars

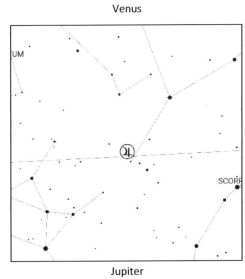

Jupiter

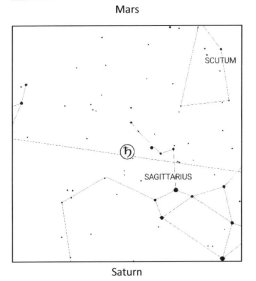

Saturn

April 1st to 10th

The Moon

 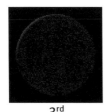   

| | 1st | | 3rd | | 5th | | 7th | | 9th | |

Date	Con	R.A.	Dec	Mag	Diam	Ill.	Elon.	Phase	Close To
1st	Aqr	22h 7m	-14° 40'	-7.8	29'	13%	39° W	-Cr	Venus
2nd	Aqr	22h 53m	-11° 2'	-7.0	30'	8%	28° W	NM	Mercury, Venus, Neptune
3rd	Aqr	23h 39m	-6° 57'	-6.1	30'	3%	18° W	NM	Mercury, Neptune
4th	Psc	0h 24m	-2° 32'	-5.1	30'	1%	7° W	NM	
5th	Cet	1h 9m	2° 3'	-4.4	30'	0%	3° E	NM	
6th	Psc	1h 56m	6° 37'	-5.4	30'	1%	14° E	NM	Uranus
7th	Ari	2h 44m	10° 58'	-6.4	31'	5%	25° E	NM	Uranus
8th	Tau	3h 34m	14° 55'	-7.3	31'	10%	37° E	NM	Mars, Pleiades
9th	Tau	4h 27m	18° 12'	-8.1	31'	17%	49° E	+Cr	Mars, Pleiades, Hyades, Aldebara
10th	Tau	5h 22m	20° 35'	-8.8	31'	26%	62° E	+Cr	Aldebaran

Mercury and Venus

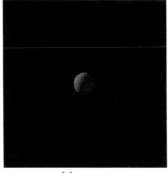

Mercury
5th

Venus
5th

Mercury

Date	Con.	R.A.	Dec.	Mag.	Diam.	Ill.	Elon.	Vis.	Rat.	Close To
1st	Aqr	23h 11m	-5° 32'	0.9	9"	30%	23° W	AM	***	Venus, Neptune
3rd	Aqr	23h 15m	-5° 35'	0.8	9"	34%	23° W	AM	***	Moon, Venus, Neptune
5th	Aqr	23h 20m	-5° 29'	0.6	9"	38%	24° W	AM	***	Venus, Neptune
7th	Aqr	23h 26m	-5° 14'	0.5	8"	41%	24° W	AM	****	Venus, Neptune
9th	Aqr	23h 32m	-4° 52'	0.5	8"	44%	25° W	AM	***	Venus, Neptune

Venus

Date	Con.	R.A.	Dec.	Mag.	Diam.	Ill.	Elon.	Vis.	Rat.	Close To
1st	Aqr	22h 35m	-9° 52'	-4.0	13"	81%	32° W	AM	**	Moon, Mercury
3rd	Aqr	22h 44m	-9° 2'	-4.0	13"	82%	31° W	AM	**	Mercury
5th	Aqr	22h 54m	-8° 12'	-4.0	13"	82%	31° W	AM	**	Mercury
7th	Aqr	23h 3m	-7° 20'	-4.0	13"	83%	30° W	AM	**	Mercury, Neptune
9th	Aqr	23h 12m	-6° 28'	-3.9	13"	83%	30° W	AM	**	Mercury, Neptune

Mars and the Outer Planets

Mars
5th

Jupiter
5th

Saturn
5th

Mars

Date	Con.	R.A.	Dec.	Mag.	Diam.	Ill.	Elon.	Vis.	Rat.	Close To
1st	Tau	3h 53m	21° 13'	1.5	5"	94%	48° E	PM	*	Pleiades, Hyades
5th	Tau	4h 4m	21° 46'	1.5	5"	94%	47° E	PM	*	Pleiades, Hyades, Aldebaran
10th	Tau	4h 18m	22° 24'	1.5	4"	94%	46° E	PM	*	Pleiades, Hyades, Aldebaran

The Outer Planets

Planet	Date	Con.	R.A.	Dec.	Mag.	Diam.	Elon.	Vis.	Rat.	Close To
Jupiter	5th	Oph	17h 35m	-22° 41'	-2.3	40"	110° W	AM	***	
Saturn	5th	Sgr	19h 26m	-21° 32'	0.6	16"	83° W	AM	***	
Uranus	5th	Ari	1h 58m	11° 33'	5.9	3"	15° E	PM	*	
Neptune	5th	Aqr	23h 14m	-5° 56'	8.0	2"	26° W	AM	*	Mercury, Venus

Highlights

Date	Time (UT)	Event
1st	N/A	Good opportunity to see Earthshine on the waning crescent Moon. (Morning sky.)
2nd	02:52	The waning crescent Moon is south of Venus. (Morning sky.)
	18:52	Mercury is 0.4° north of Neptune. (Morning sky.)
	22:46	The waning crescent Moon is south of Neptune. (Morning sky.)
	22:54	The waning crescent Moon is south of Mercury. (Morning sky.)
5th	08:51	New Moon. (Not visible.)
8th	01:50	Dwarf planet Ceres is stationary prior to beginning retrograde motion. (Morning sky.)
	18:49	The waxing crescent Moon is south of the Pleiades star cluster. (Taurus, evening sky.)
	N/A	Good opportunity to see Earthshine on the waxing crescent Moon. (Evening sky.)
9th	05:49	The waxing crescent Moon is south of Mars. (Evening sky.)
	15:52	The waxing crescent Moon is north of the bright star Aldebaran. (Taurus, evening sky.)
10th	03:42	Venus is 0.3° south of Neptune. (Morning sky.)
	16:15	Jupiter is stationary prior to beginning retrograde motion. (Morning sky.)

Planet Locations – April 5th

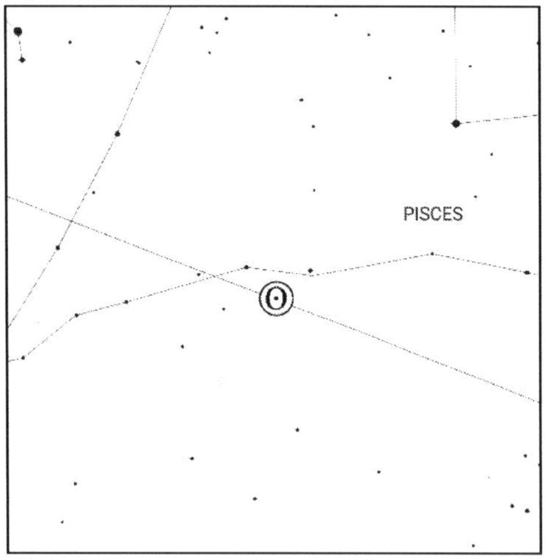

| ☉ Sun | ☿ Mercury | ♀ Venus | ♂ Mars | ♃ Jupiter | ♄ Saturn |

PISCES

Sun

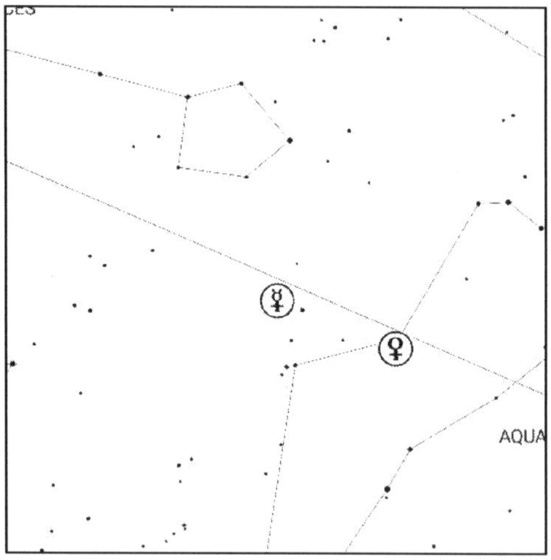

Mercury

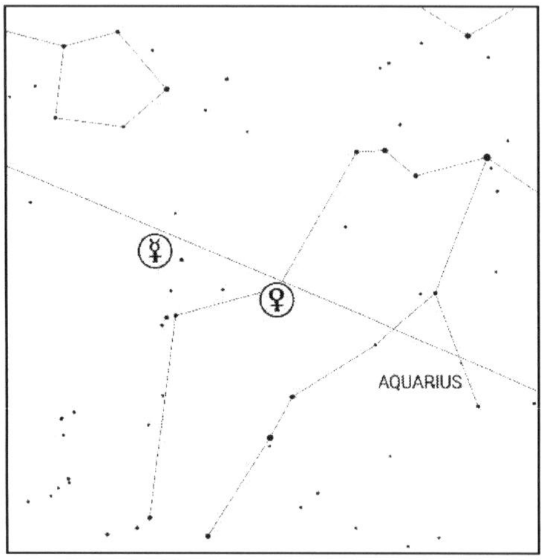

AQUA

AQUARIUS

Venus

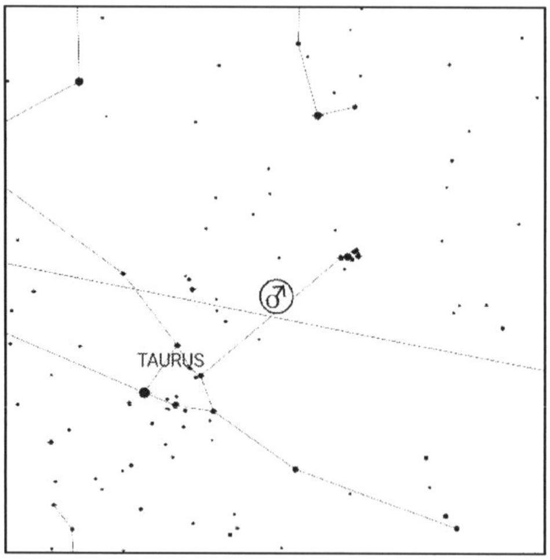

TAURUS

Mars

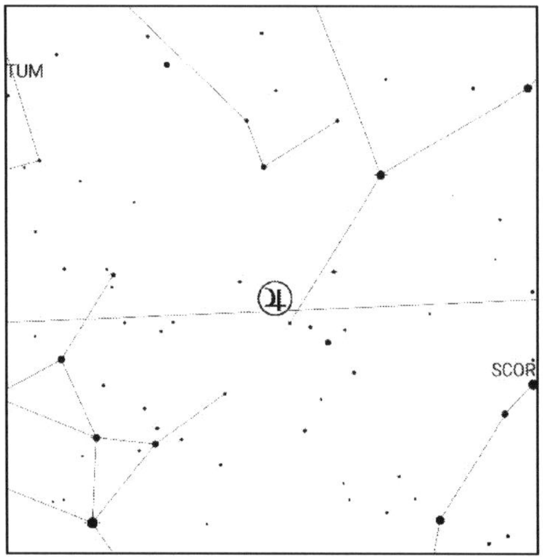

TUM

SCOR

Jupiter

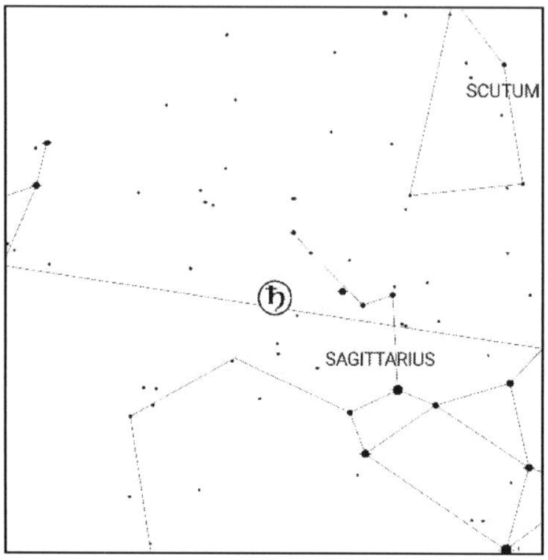

SCUTUM

SAGITTARIUS

Saturn

The Moon

| | 11th | | 13th | | 15th | | 17th | | 19th |

Date	Con	R.A.	Dec	Mag	Diam	Ill.	Elon.	Phase	Close To
11th	Gem	6h 20m	21° 51'	-9.4	32'	36%	75° E	+Cr	
12th	Gem	7h 19m	21° 51'	-10.0	32'	47%	89° E	FQ	
13th	Cnc	8h 19m	20° 30'	-10.5	32'	58%	103° E	FQ	Praesepe
14th	Cnc	9h 19m	17° 51'	-10.9	33'	69%	117° E	+G	Praesepe
15th	Leo	10h 17m	14° 4'	-11.3	33'	79%	131° E	+G	Regulus
16th	Leo	11h 13m	9° 23'	-11.7	33'	88%	144° E	+G	
17th	Vir	12h 8m	4° 8'	-12.0	33'	94%	157° E	+G	
18th	Vir	13h 3m	-1° 20'	-12.4	33'	99%	170° E	FM	Spica
19th	Vir	13h 57m	-6° 41'	-12.6	32'	100%	178° W	FM	Spica
20th	Lib	14h 51m	-11° 34'	-12.4	32'	98%	165° W	FM	

Mercury and Venus

Mercury
15th

Venus
15th

Mercury

Date	Con.	R.A.	Dec.	Mag.	Diam.	Ill.	Elon.	Vis.	Rat.	Close To
11th	Aqr	23h 39m	-4° 22'	0.4	8"	48%	25° W	AM	***	Venus
13th	Aqr	23h 47m	-3° 45'	0.3	8"	51%	25° W	AM	***	Venus
15th	Psc	23h 56m	-3° 2'	0.3	7"	54%	24° W	AM	***	Venus
17th	Psc	0h 5m	-2° 13'	0.2	7"	56%	24° W	AM	***	Venus
19th	Psc	0h 14m	-1° 18'	0.1	7"	59%	23° W	AM	***	Venus

Venus

Date	Con.	R.A.	Dec.	Mag.	Diam.	Ill.	Elon.	Vis.	Rat.	Close To
11th	Aqr	23h 21m	-5° 35'	-3.9	12"	84%	29° W	AM	**	Mercury, Neptune
13th	Aqr	23h 30m	-4° 41'	-3.9	12"	84%	29° W	AM	**	Mercury, Neptune
15th	Aqr	23h 39m	-3° 47'	-3.9	12"	85%	29° W	AM	*	Mercury
17th	Psc	23h 48m	-2° 52'	-3.9	12"	85%	28° W	AM	*	Mercury
19th	Psc	23h 57m	-1° 57'	-3.9	12"	86%	28° W	AM	*	Mercury

Mars and the Outer Planets

Mars
15th

Jupiter
15th

Saturn
15th

Mars

Date	Con.	R.A.	Dec.	Mag.	Diam.	Ill.	Elon.	Vis.	Rat.	Close To
11th	Tau	4h 21m	22° 31'	1.5	4"	94%	46° E	PM	*	Pleiades, Hyades, Aldebaran
15th	Tau	4h 33m	22° 57'	1.5	4"	95%	45° E	PM	*	Hyades, Aldebaran
20th	Tau	4h 47m	23° 25'	1.6	4"	95%	44° E	PM	*	Hyades, Aldebaran

The Outer Planets

Planet	Date	Con.	R.A.	Dec.	Mag.	Diam.	Elon.	Vis.	Rat.	Close To
Jupiter	15th	Oph	17h 35m	-22° 41'	-2.3	42"	120° W	AM	***	
Saturn	15th	Sgr	19h 27m	-21° 30'	0.5	17"	91° W	AM	***	
Uranus	15th	Ari	2h 0m	11° 45'	5.9	3"	7° E	NV	N/A	
Neptune	15th	Aqr	23h 15m	-5° 48'	7.9	2"	34° W	AM	*	Venus

Highlights

Date	Time (UT)	Event
11th	19:29	Mercury is at greatest western elongation. (Morning sky.)
12th	19:06	First Quarter Moon. (Evening sky.)
13th	20:37	The just-past first quarter Moon is south of the Praesepe star cluster. (Cancer, evening sky.)
15th	08:24	The waxing gibbous Moon is north of the bright star Regulus. (Leo, evening sky.)
16th	12:11	Mars is 6.5° north of the bright star Aldebaran. (Taurus, evening sky.)
18th	20:54	The nearly full Moon is north of the bright star Spica. (Virgo, evening sky.)
20th	11:13	Full Moon. (Visible all night.)

Planet Locations – April 15th

| ☉ Sun | ☿ Mercury | ♀ Venus | ♂ Mars | ♃ Jupiter | ♄ Saturn |

Sun

Mercury

Venus

Mars

Jupiter

Saturn

April 21st – 30th

The Moon

21st   23rd   25th   27th   29th

Date	Con	R.A.	Dec	Mag	Diam	Ill.	Elon.	Phase	Close To
21st	Lib	15h 45m	-15° 43'	-12.0	32'	95%	153° W	-G	Antares
22nd	Oph	16h 40m	-18° 56'	-11.7	31'	89%	140° W	-G	Antares
23rd	Oph	17h 35m	-21° 3'	-11.4	31'	81%	127° W	-G	Jupiter
24th	Sgr	18h 29m	-22° 3'	-11.0	30'	73%	114° W	-G	
25th	Sgr	19h 23m	-21° 56'	-10.7	30'	64%	102° W	LQ	Saturn
26th	Cap	20h 14m	-20° 47'	-10.3	30'	54%	90° W	LQ	Saturn
27th	Cap	21h 4m	-18° 43'	-9.9	30'	45%	78° W	LQ	
28th	Cap	21h 53m	-15° 53'	-9.4	30'	35%	67° W	-Cr	
29th	Aqr	22h 39m	-12° 24'	-8.9	30'	27%	57° W	-Cr	Neptune
30th	Aqr	23h 25m	-8° 24'	-8.3	30'	19%	46° W	-Cr	Neptune

Mercury and Venus

Mercury
25th

Venus
25th

Mercury

Date	Con.	R.A.	Dec.	Mag.	Diam.	Ill.	Elon.	Vis.	Rat.	Close To
21st	Psc	0h 24m	0° 18'	0.1	7"	62%	23° W	AM	***	Venus
23rd	Cet	0h 34m	0° 47'	0.0	6"	65%	22° W	AM	***	Venus
25th	Cet	0h 45m	1° 56'	-0.1	6"	67%	21° W	AM	***	Venus
27th	Psc	0h 56m	3° 10'	-0.2	6"	70%	20° W	AM	***	Venus
29th	Psc	1h 8m	4° 28'	-0.3	6"	73%	19° W	AM	***	Venus

Venus

Date	Con.	R.A.	Dec.	Mag.	Diam.	Ill.	Elon.	Vis.	Rat.	Close To
21st	Psc	0h 6m	-1° 1'	-3.9	12"	86%	27° W	AM	*	Mercury
23rd	Psc	0h 15m	0° 5'	-3.9	12"	87%	27° W	AM	*	Mercury
25th	Psc	0h 24m	0° 51'	-3.9	12"	87%	27° W	AM	*	Mercury
27th	Cet	0h 32m	1° 47'	-3.9	12"	87%	26° W	AM	*	Mercury
29th	Cet	0h 41m	2° 43'	-3.9	12"	88%	26° W	AM	*	Mercury

Mars and the Outer Planets

Mars
25th

Jupiter
25th

Saturn
25th

Mars

Date	Con.	R.A.	Dec.	Mag.	Diam.	Ill.	Elon.	Vis.	Rat.	Close To
21st	Tau	4h 49m	23° 30'	1.6	4"	95%	44° E	PM	*	Hyades, Aldebaran
25th	Tau	5h 1m	23° 48'	1.6	4"	95%	43° E	PM	*	Hyades, Aldebaran
30th	Tau	5h 15m	24° 7'	1.6	4"	96%	41° E	PM	*	

The Outer Planets

Planet	Date	Con.	R.A.	Dec.	Mag.	Diam.	Elon.	Vis.	Rat.	Close To
Jupiter	25th	Oph	17h 34m	-22° 40'	-2.4	43"	129° W	AM	****	
Saturn	25th	Sgr	19h 28m	-21° 29'	0.5	17"	101° W	AM	***	Moon
Uranus	25th	Ari	2h 2m	11° 56'	5.9	3"	2° W	NV	N/A	
Neptune	25th	Aqr	23h 16m	-5° 42'	7.9	2"	43° W	AM	**	

Highlights

Date	Time (UT)	Event
22nd	08:24	The waning gibbous Moon is north of the bright star Antares. (Scorpius, morning sky.)
	N/A	The Lyrid meteor shower is at its maximum. (ZHR: 18)
23rd	03:46	Uranus is in conjunction with the Sun. (Not visible.)
	12:29	The waning gibbous Moon is north of Jupiter. (Morning sky.)
24th	18:21	Dwarf planet Pluto is stationary prior to beginning retrograde motion. (Morning sky.)
25th	15:10	The nearly last quarter Moon is south of Saturn. (Morning sky.)
	18:53	The nearly last quarter Moon is south of dwarf planet Pluto. (Morning sky.)
26th	22:19	Last Quarter Moon. (Morning sky.)
30th	01:27	Saturn is stationary prior to beginning retrograde motion. (Morning sky.)
	07:18	The waning crescent Moon is south of Neptune. (Morning sky.)

Planet Locations – April 25th

☉	☿	♀	♂	♃	♄
Sun	Mercury	Venus	Mars	Jupiter	Saturn

Sun

Mercury

Venus

Mars

Jupiter

Saturn

The Moon

| | 1st | | 3rd | | 5th | | 7th | | 9th |

Date	Con	R.A.	Dec	Mag	Diam	Ill.	Elon.	Phase		Close To
1st	Psc	0h 10m	-4° 2'	-7.6	30'	12%	36° W	NM		Venus
2nd	Cet	0h 55m	0° 34'	-6.7	30'	6%	25° W	NM		Mercury, Venus
3rd	Psc	1h 42m	5° 14'	-5.8	30'	2%	15° W	NM		Mercury, Venus, Uranus
4th	Cet	2h 30m	9° 46'	-4.7	31'	0%	4° W	NM		Uranus
5th	Ari	3h 20m	13° 57'	-4.7	31'	0%	8° E	NM		Pleiades
6th	Tau	4h 13m	17° 32'	-5.9	31'	3%	20° E	NM		Pleiades, Hyades, Aldebaran
7th	Tau	5h 9m	20° 16'	-6.9	31'	7%	33° E	NM		Mars, Hyades, Aldebaran
8th	Gem	6h 7m	21° 53'	-7.8	32'	14%	47° E	+Cr		Mars
9th	Gem	7h 6m	22° 12'	-8.6	32'	23%	61° E	+Cr		
10th	Cnc	8h 6m	21° 10'	-9.3	32'	33%	75° E	+Cr		Praesepe

Mercury and Venus

Mercury
5th

Venus
5th

Mercury

Date	Con.	R.A.	Dec.	Mag.	Diam.	Ill.	Elon.	Vis.	Rat.		Close To
1st	Psc	1h 20m	5° 49'	-0.4	6"	76%	18° W	AM	**		Venus
3rd	Psc	1h 32m	7° 13'	-0.5	6"	79%	17° W	AM	**		Moon, Venus
5th	Psc	1h 45m	8° 41'	-0.6	6"	82%	16° W	AM	**		Venus, Uranus
7th	Psc	1h 59m	10° 10'	-0.7	5"	85%	14° W	NV	N/A		Uranus
9th	Ari	2h 13m	11° 41'	-0.9	5"	88%	13° W	NV	N/A		Uranus

Venus

Date	Con.	R.A.	Dec.	Mag.	Diam.	Ill.	Elon.	Vis.	Rat.	Close To
1st	Psc	0h 50m	3° 39'	-3.9	12"	88%	26° W	AM	*	Moon, Mercury
3rd	Psc	0h 59m	4° 35'	-3.9	11"	89%	25° W	AM	*	Moon, Mercury
5th	Psc	1h 8m	5° 30'	-3.9	11"	89%	25° W	AM	*	Mercury
7th	Psc	1h 17m	6° 26'	-3.9	11"	89%	25° W	AM	*	
9th	Psc	1h 26m	7° 20'	-3.9	11"	90%	24° W	AM	*	

Mars and the Outer Planets

Mars
5th

Jupiter
5th

Saturn
5th

Mars

Date	Con.	R.A.	Dec.	Mag.	Diam.	Ill.	Elon.	Vis.	Rat.	Close To
1st	Tau	5h 18m	24° 10'	1.6	4"	96%	41° E	PM	*	
5th	Tau	5h 29m	24° 21'	1.7	4"	96%	40° E	PM	*	
10th	Tau	5h 43m	24° 30'	1.7	4"	96%	39° E	PM	*	

The Outer Planets

Planet	Date	Con.	R.A.	Dec.	Mag.	Diam.	Elon.	Vis.	Rat.	Close To
Jupiter	5th	Oph	17h 31m	-22° 39'	-2.5	44"	139° W	AM	****	
Saturn	5th	Sgr	19h 28m	-21° 29'	0.4	17"	110° W	AM	***	
Uranus	5th	Ari	2h 4m	12° 8'	5.9	3"	11° W	NV	N/A	Mercury
Neptune	5th	Aqr	23h 17m	-5° 36'	7.9	2"	53° W	AM	**	

Highlights

Date	Time (UT)	Event
2nd	12:35	The waning crescent Moon is south of Venus. (Morning sky.)
3rd	04:56	The waning crescent Moon is south of Mercury. (Morning sky.)
4th	22:46	New Moon. (Not visible.)
6th	22:43	The waxing crescent Moon is north of the bright star Aldebaran. (Taurus, evening sky.)
	N/A	The Eta Aquariid meteor shower is at its maximum. (ZHR: 70)
7th	N/A	Good opportunity to see Earthshine on the waxing crescent Moon. (Evening sky.)
8th	00:18	The waxing crescent Moon is south of Mars. (Evening sky.)
9th	N/	The Eta Lyrid meteor shower is at its maximum. (ZHR: 3)

Planet Locations – May 5th

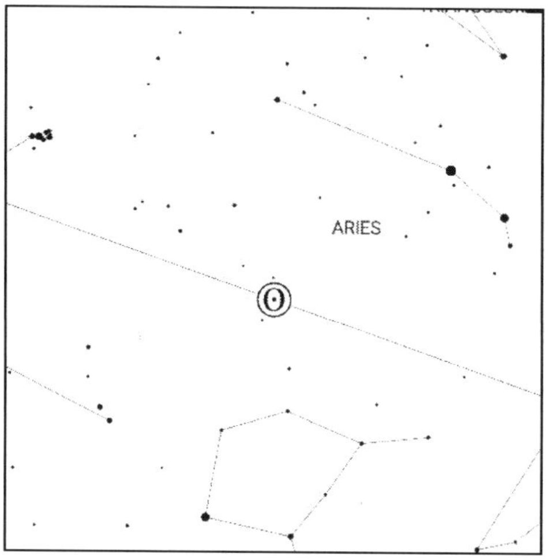

Sun 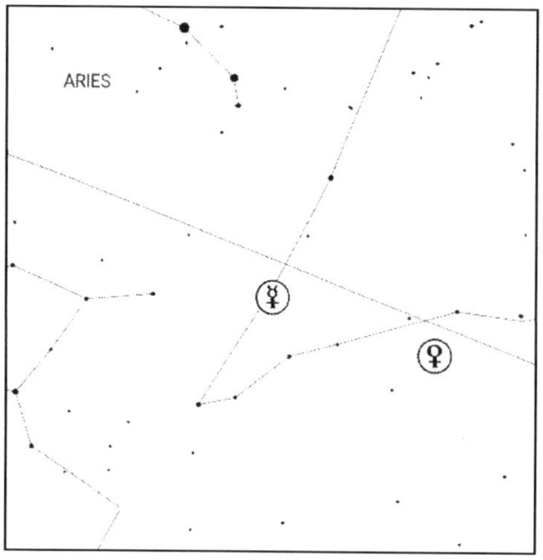   Mercury 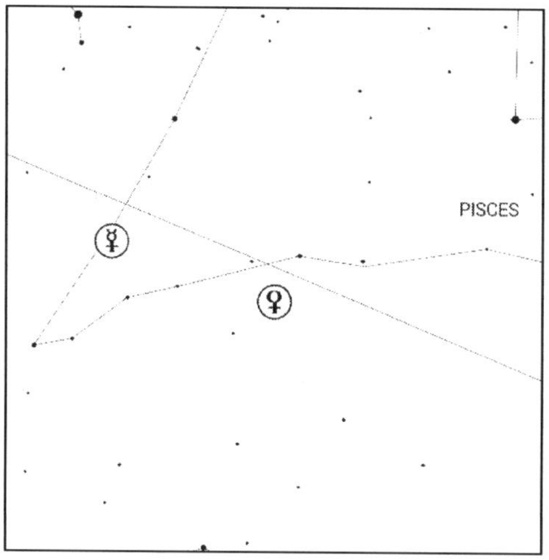   Venus 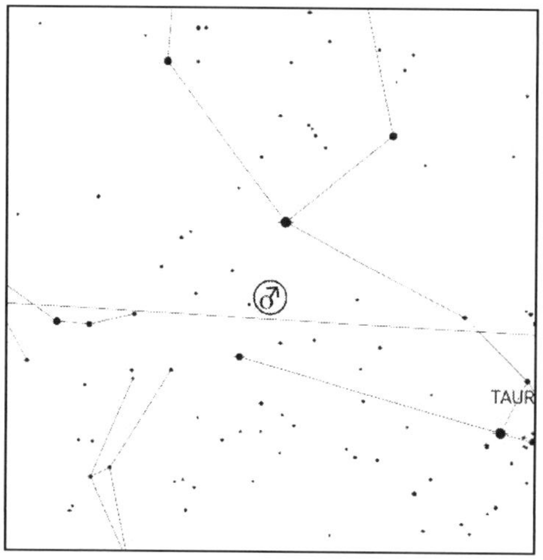   Mars 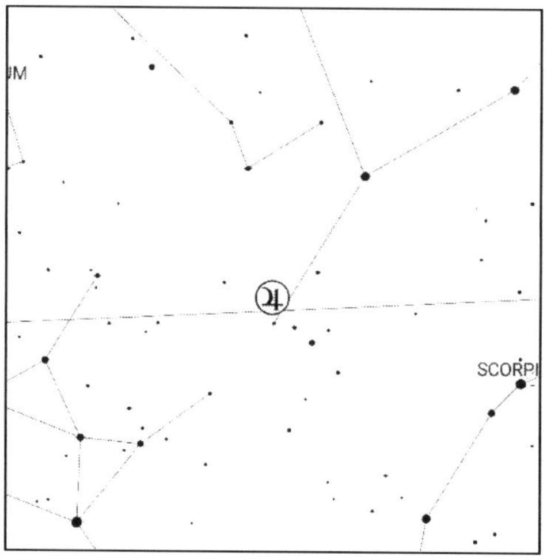   Jupiter   Saturn

Sun

Mercury

Venus

Mars

Jupiter

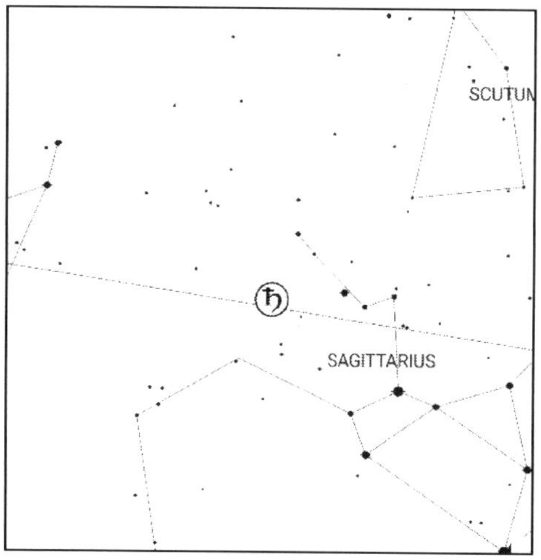

Saturn

May 11th – 20th

The Moon

| | 11th | | 13th | | 15th | | 17th | | 19th |

Date	Con	R.A.	Dec	Mag	Diam	Ill.	Elon.	Phase	Close To
11th	Cnc	9h 5m	18° 49'	-9.8	32'	44%	88° E	FQ	Praesepe
12th	Leo	10h 2m	15° 19'	-10.3	32'	55%	102° E	FQ	Regulus
13th	Leo	10h 58m	10° 56'	-10.8	32'	66%	115° E	+G	
14th	Vir	11h 52m	5° 55'	-11.2	32'	77%	127° E	+G	
15th	Vir	12h 45m	0° 35'	-11.6	32'	86%	139° E	+G	Spica
16th	Vir	13h 38m	-4° 45'	-11.9	32'	93%	151° E	+G	Spica
17th	Lib	14h 30m	-9° 48'	-12.3	32'	97%	164° E	FM	
18th	Lib	15h 24m	-14° 15'	-12.6	32'	100%	176° E	FM	
19th	Sco	16h 18m	-17° 53'	-12.5	31'	99%	171° W	FM	Antares
20th	Oph	17h 14m	-20° 30'	-12.2	31'	97%	158° W	FM	Jupiter, Antares

Mercury and Venus

Mercury
15th

Venus
15th

Mercury

Date	Con.	R.A.	Dec.	Mag.	Diam.	Ill.	Elon.	Vis.	Rat.	Close To
11th	Ari	2h 28m	13° 14'	-1.1	5"	91%	11° W	NV	N/A	
13th	Ari	2h 43m	14° 46'	-1.3	5"	94%	9° W	NV	N/A	
15th	Ari	2h 59m	16° 17'	-1.5	5"	96%	7° W	NV	N/A	
17th	Ari	3h 16m	17° 45'	-1.8	5"	98%	5° W	NV	N/A	Pleiades
19th	Tau	3h 33m	19° 10'	-2.0	5"	100%	3° W	NV	N/A	Pleiades

Venus

Date	Con.	R.A.	Dec.	Mag.	Diam.	Ill.	Elon.	Vis.	Rat.	Close To
11th	Psc	1h 36m	8° 14'	-3.9	11"	90%	24° W	AM	*	
13th	Psc	1h 45m	9° 8'	-3.9	11"	91%	24° W	AM	*	
15th	Psc	1h 54m	10° 0'	-3.9	11"	91%	23° W	AM	*	Uranus
17th	Ari	2h 3m	10° 52'	-3.9	11"	91%	23° W	AM	*	Uranus
19th	Ari	2h 12m	11° 43'	-3.9	11"	92%	23° W	AM	*	Uranus

Mars and the Outer Planets

Mars
15th

Jupiter
15th

Saturn
15th

Mars

Date	Con.	R.A.	Dec.	Mag.	Diam.	Ill.	Elon.	Vis.	Rat.	Close To
11th	Tau	5h 46m	24° 31'	1.7	4"	96%	39° E	PM	*	
15th	Tau	5h 58m	24° 33'	1.7	4"	97%	38° E	PM	*	
20th	Gem	6h 12m	24° 32'	1.7	4"	97%	36° E	PM	*	

The Outer Planets

Planet	Date	Con.	R.A.	Dec.	Mag.	Diam.	Elon.	Vis.	Rat.	Close To
Jupiter	15th	Oph	17h 27m	-22° 36'	-2.5	45"	150° W	AM	****	
Saturn	15th	Sgr	19h 27m	-21° 31'	0.4	18"	120° W	AM	****	
Uranus	15th	Ari	2h 6m	12° 20'	5.9	3"	20° W	AM	*	Venus
Neptune	15th	Aqr	23h 18m	-5° 31'	7.9	2"	62° W	AM	**	

Highlights

Date	Time (UT)	Event
11th	01:13	First Quarter Moon. (Evening sky.)
	13:04	The first quarter Moon is north of the bright star Regulus. (Leo, evening sky.)
16th	07:00	The waxing gibbous Moon is north of the bright star Spica. (Virgo, evening sky.)
18th	08:09	Venus is 1.2° south of Uranus. (Morning sky.)
	21:12	Full Moon. (Visible all night.)
19th	15:39	The just-past full Moon is north of the bright star Antares. (Scorpius, visible all night.)
20th	16:07	The waning gibbous Moon is north of Jupiter. (Morning sky.)

Planet Locations – May 15th

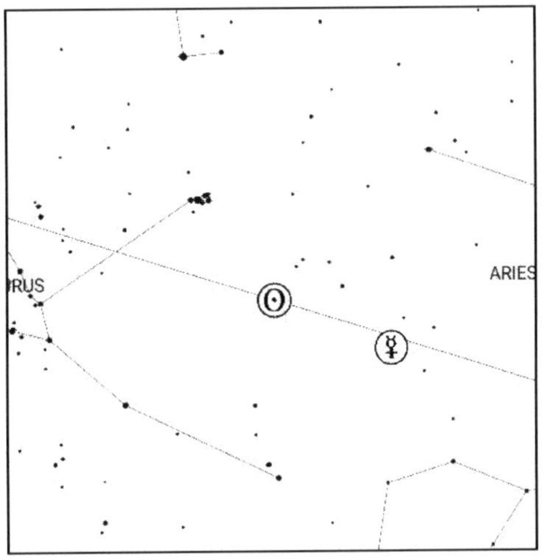

Sun	Mercury	Venus	Mars	Jupiter	Saturn

Sun

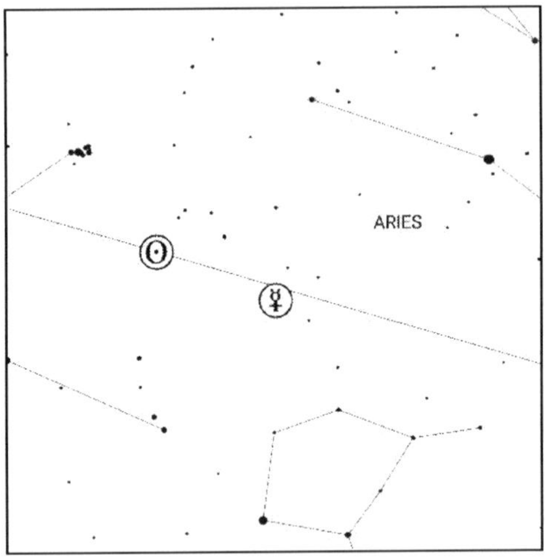

ARIES

TAURUS

Mercury

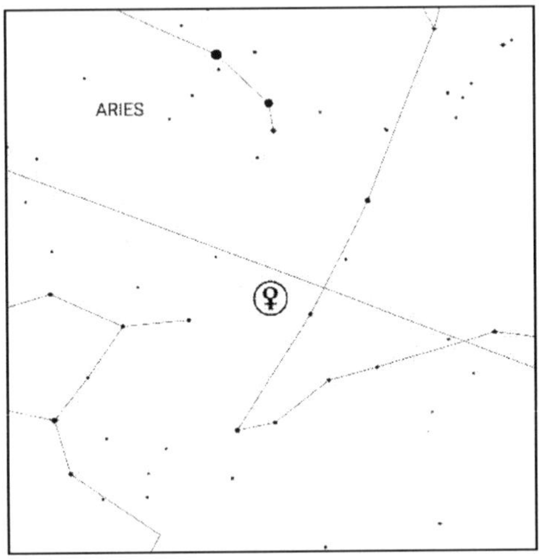

ARIES

Venus

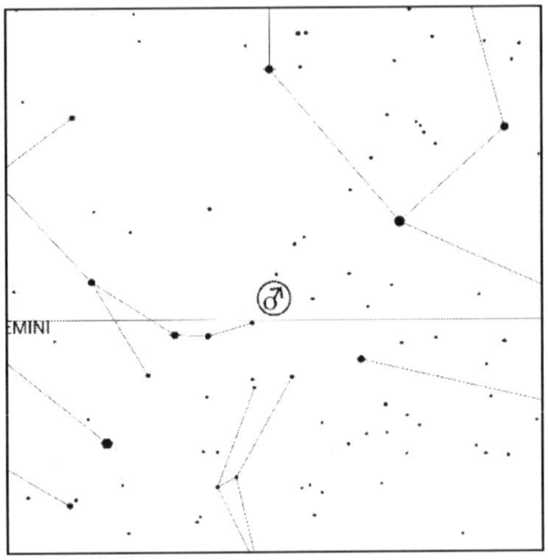

ARIES

Mars

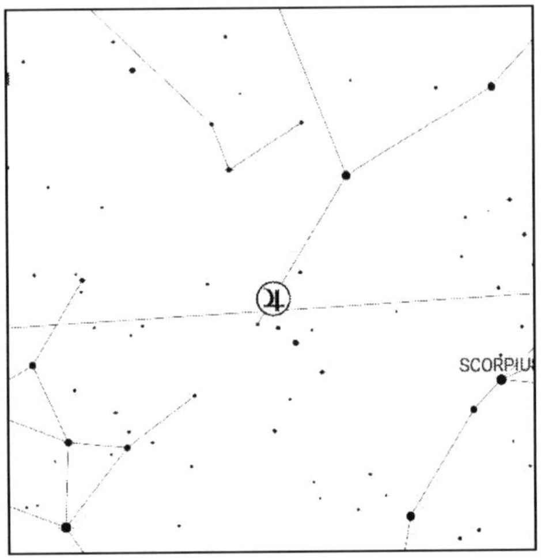

GEMINI

Jupiter

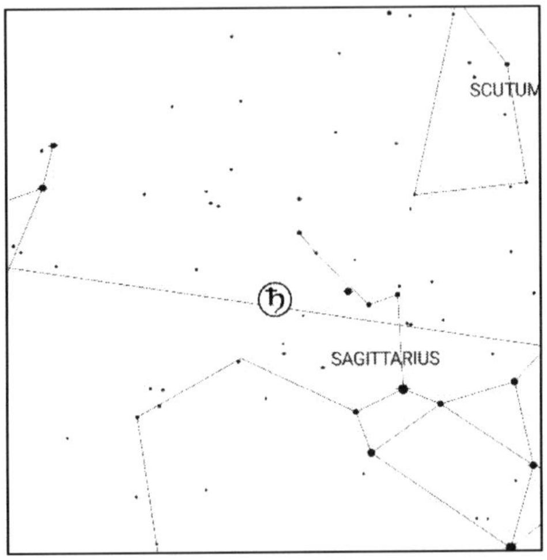

SCORPIUS

Saturn

SCUTUM

SAGITTARIUS

The Moon

| | 21ˢᵗ | | 23ʳᵈ | | 25ᵗʰ | | 27ᵗʰ | | 29ᵗʰ | | 31ˢᵗ |

Date	Con	R.A.	Dec	Mag	Diam	Ill.	Elon.	Phase	Close To
21st	Sgr	18h 9m	-21° 58'	-11.9	31'	92%	146° W	-G	Jupiter
22nd	Sgr	19h 3m	-22° 17'	-11.6	30'	86%	133° W	-G	Saturn
23rd	Sgr	19h 57m	-21° 30'	-11.3	30'	79%	121° W	-G	Saturn
24th	Cap	20h 48m	-19° 44'	-10.9	30'	70%	109° W	-G	
25th	Cap	21h 37m	-17° 7'	-10.6	30'	61%	98° W	LQ	
26th	Aqr	22h 24m	-13° 49'	-10.2	30'	52%	87° W	LQ	
27th	Aqr	23h 10m	-9° 58'	-9.8	30'	42%	76° W	LQ	Neptune
28th	Aqr	23h 55m	-5° 43'	-9.3	30'	33%	66° W	-Cr	Neptune
29th	Cet	0h 40m	-1° 11'	-8.7	30'	24%	56° W	-Cr	
30th	Psc	1h 26m	3° 29'	-8.1	30'	16%	46° W	-Cr	Uranus
31st	Cet	2h 13m	8° 7'	-7.3	31'	10%	35° W	NM	Uranus

Mercury and Venus

Mercury
25ᵗʰ

Venus
25ᵗʰ

Mercury

Date	Con.	R.A.	Dec.	Mag.	Diam.	Ill.	Elon.	Vis.	Rat.	Close To
21st	Tau	3h 51m	20° 29'	-2.3	5"	100%	0° W	NV	N/A	Pleiades, Hyades
23rd	Tau	4h 9m	21° 41'	-2.1	5"	99%	2° E	NV	N/A	Pleiades, Hyades, Aldebara
25th	Tau	4h 27m	22° 45'	-1.8	5"	98%	5° E	NV	N/A	Hyades, Aldebaran
27th	Tau	4h 46m	23° 39'	-1.6	5"	95%	8° E	NV	N/A	Hyades, Aldebaran
29th	Tau	5h 4m	24° 22'	-1.4	5"	92%	10° E	NV	N/A	Hyades, Aldebaran
31st	Tau	5h 22m	24° 55'	-1.2	5"	88%	13° E	NV	N/A	

Venus

Date	Con.	R.A.	Dec.	Mag.	Diam.	Ill.	Elon.	Vis.	Rat.	Close To
21st	Ari	2h 22m	12° 33'	-3.9	11"	92%	22° W	AM	*	Uranus
23rd	Ari	2h 31m	13° 21'	-3.9	11"	92%	22° W	AM	*	
25th	Ari	2h 41m	14° 9'	-3.9	11"	93%	22° W	AM	*	
27th	Ari	2h 50m	14° 55'	-3.9	11"	93%	21° W	AM	*	
29th	Ari	3h 0m	15° 40'	-3.9	11"	93%	21° W	AM	*	
31st	Ari	3h 9m	16° 23'	-3.9	11"	94%	21° W	AM	*	Pleiades

Mars and the Outer Planets

Mars
25th

Jupiter
25th

Saturn
25th

Mars

Date	Con.	R.A.	Dec.	Mag.	Diam.	Ill.	Elon.	Vis.	Rat.	Close To
21st	Gem	6h 15m	24° 32'	1.7	4"	97%	36° E	PM	*	
25th	Gem	6h 26m	24° 26'	1.7	4"	97%	35° E	PM	*	
31st	Gem	6h 43m	24° 13'	1.8	4"	97%	33° E	PM	*	

The Outer Planets

Planet	Date	Con.	R.A.	Dec.	Mag.	Diam.	Elon.	Vis.	Rat.	Close To
Jupiter	25th	Oph	17h 23m	-22° 33'	-2.6	46"	161° W	AM	****	
Saturn	25th	Sgr	19h 26m	-21° 35'	0.3	18"	130° W	AM	****	
Uranus	25th	Ari	2h 8m	12° 30'	5.9	3"	30° W	AM	*	Venus
Neptune	25th	Aqr	23h 19m	-5° 27'	7.9	2"	72° W	AM	**	

Highlights

Date	Time (UT)	Event
21st	12:54	Mercury is at superior conjunction with the Sun. (Not visible.)
22nd	21:01	The waning gibbous Moon is south of Saturn. (Morning sky.)
23rd	03;23	The waning gibbous Moon is south of dwarf planet Pluto. (Morning sky.)
24th	N/A	The May Camelopardalid meteor shower is at its maximum. (ZHR: Variable.)
26th	16:34	Last Quarter Moon. (Morning sky.)
27th	17:18	The waning crescent Moon is south of Neptune. (Morning sky.)
29th	04:10	Dwarf planet Ceres is at opposition. (Visible all night.)
30th	N/A	Good opportunity to see Earthshine on the waning crescent Moon. (Morning sky.)
31st	10:55	The waning crescent Moon is south of Uranus. (Morning sky.)

Planet Locations – May 25th

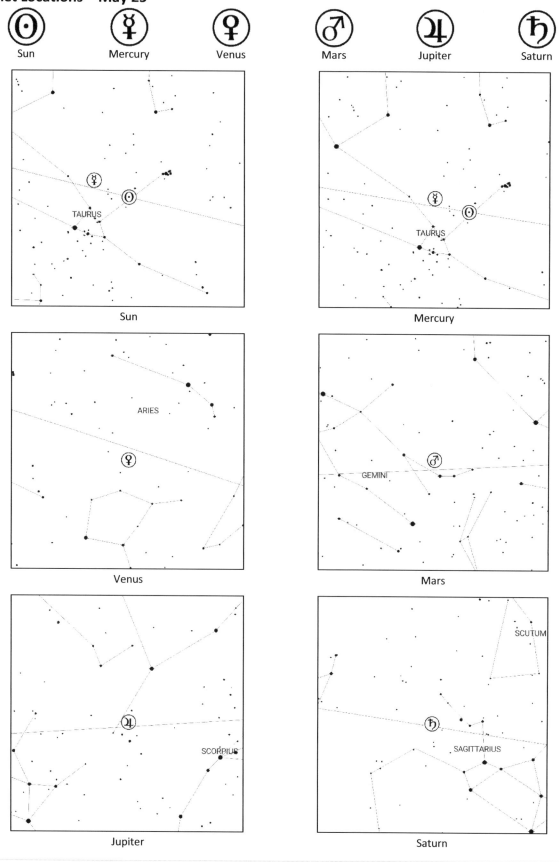

June 1st to 10th

The Moon

  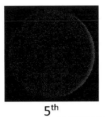  

| 1st | 3rd | 5th | 7th | 9th |

Date	Con	R.A.	Dec	Mag	Diam	Ill.	Elon.	Phase	Close To
1st	Ari	3h 2m	12° 30'	-6.3	31'	4%	23° W	NM	Venus, Pleiades
2nd	Tau	3h 55m	16° 24'	-5.2	31'	1%	11° W	NM	Venus, Pleiades, Hyades, Aldebar
3rd	Tau	4h 50m	19° 32'	-4.2	32'	0%	1° E	NM	Hyades, Aldebaran
4th	Tau	5h 49m	21° 36'	-5.4	32'	1%	15° E	NM	Mercury
5th	Gem	6h 49m	22° 22'	-6.6	32'	5%	29° E	NM	Mercury, Mars
6th	Gem	7h 51m	21° 42'	-7.6	32'	12%	44° E	NM	
7th	Cnc	8h 51m	19° 39'	-8.4	32'	20%	58° E	+Cr	Praesepe
8th	Leo	9h 50m	16° 22'	-9.1	32'	30%	71° E	+Cr	Regulus
9th	Leo	10h 46m	12° 8'	-9.7	32'	41%	84° E	FQ	Regulus
10th	Vir	11h 40m	7° 15'	-10.2	32'	53%	97° E	FQ	

Mercury and Venus

Mercury
5th

Venus
5th

Mercury

Date	Con.	R.A.	Dec.	Mag.	Diam.	Ill.	Elon.	Vis.	Rat.	Close To
1st	Tau	5h 31m	25° 7'	-1.1	5"	86%	14° E	NV	N/A	
3rd	Tau	5h 48m	25° 23'	-0.9	6"	81%	16° E	PM	**	
5th	Gem	6h 5m	25° 30'	-0.7	6"	77%	18° E	PM	**	Moon
7th	Gem	6h 21m	25° 27'	-0.6	6"	72%	20° E	PM	***	
9th	Gem	6h 36m	25° 16'	-0.4	6"	68%	22° E	PM	***	Mars

Venus

Date	Con.	R.A.	Dec.	Mag.	Diam.	Ill.	Elon.	Vis.	Rat.	Close To
1st	Ari	3h 14m	16° 44'	-3.9	10"	94%	20° W	AM	*	Moon, Pleiades
3rd	Ari	3h 24m	17° 25'	-3.9	10"	94%	20° W	AM	*	Pleiades
5th	Tau	3h 34m	18° 4'	-3.9	10"	95%	20° W	AM	*	Pleiades
7th	Tau	3h 44m	18° 41'	-3.9	10"	95%	19° W	AM	*	Pleiades
9th	Tau	3h 54m	19° 16'	-3.9	10"	95%	19° W	AM	*	Pleiades, Hyades

Mars and the Outer Planets

Mars
5th

Jupiter
5th

Saturn
5th

Mars

Date	Con.	R.A.	Dec.	Mag.	Diam.	Ill.	Elon.	Vis.	Rat.	Close To
1st	Gem	6h 46m	24° 10'	1.8	4"	98%	32° E	PM	*	
5th	Gem	6h 57m	23° 57'	1.8	4"	98%	31° E	PM	*	Moon
10th	Gem	7h 11m	23° 36'	1.8	4"	98%	29° E	PM	*	Mercury

The Outer Planets

Planet	Date	Con.	R.A.	Dec.	Mag.	Diam.	Elon.	Vis.	Rat.	Close To
Jupiter	5th	Oph	17h 17m	-22° 29'	-2.6	46"	174° W	AN	*****	
Saturn	5th	Sgr	19h 24m	-21° 40'	0.3	18"	142° W	AM	****	
Uranus	5th	Ari	2h 11m	12° 41'	5.9	3"	40° W	AM	*	
Neptune	5th	Aqr	23h 19m	-5° 25'	7.9	2"	83° W	AM	***	

Highlights

Date	Time (UT)	Event
1st	19:20	The waning crescent Moon is south of Venus. (Morning sky.)
3rd	10:03	New Moon. (Not visible.)
4th	16:52	The waxing crescent Moon is south of Mercury.. (Evening sky.)
5th	15:34	The waxing crescent Moon is south of Mars. (Evening sky.)
6th	N/A	Good opportunity to see Earthshine on the waxing crescent Moon. (Evening sky.)
7th	06:26	The waxing crescent Moon is south of the Praesepe star cluster. (Cancer, evening sky.)
8th	00:22	Venus is 5.3° south of the Pleiades star cluster. (Morning sky.)
	20:32	The waxing crescent Moon is north of the bright star Regulus. (Leo, evening sky.)
10th	06:00	First Quarter Moon. (Evening sky.)
	16:31	Jupiter is at opposition. (Visible all night.)

Planet Locations – June 5th

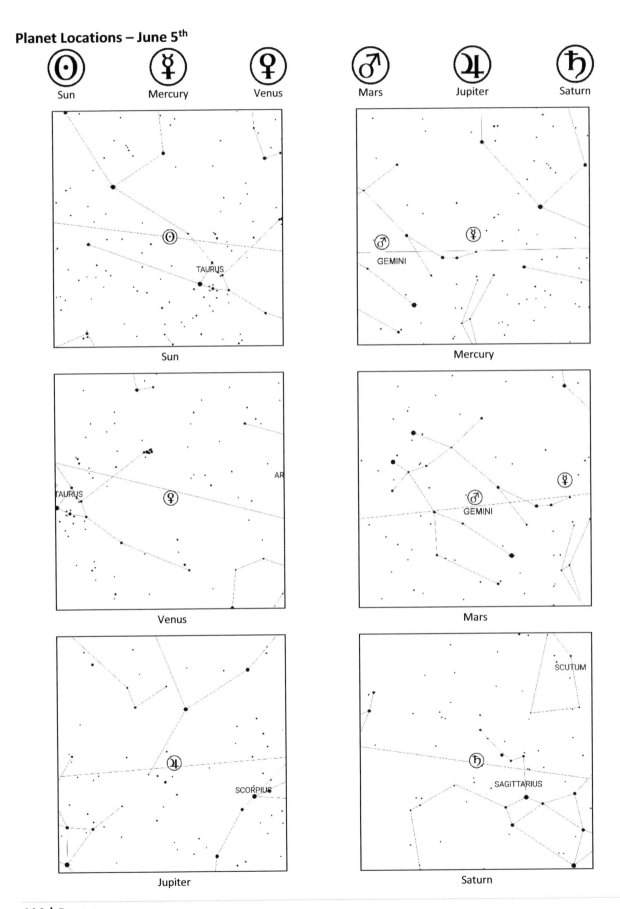

Sun   Mercury   Venus   Mars   Jupiter   Saturn

Sun

Mercury

Venus

Mars

Jupiter

Saturn

The Moon

| | 11th | | 13th | | 15th | | 17th | | 19th |

Date	Con	R.A.	Dec	Mag	Diam	Ill.	Elon.	Phase	Close To
11th	Vir	12h 32m	2° 2'	-10.7	32'	64%	109° E	FQ	
12th	Vir	13h 24m	-3° 15'	-11.1	32'	75%	121° E	+G	Spica
13th	Vir	14h 15m	-8° 19'	-11.5	32'	84%	132° E	+G	
14th	Lib	15h 8m	-12° 55'	-11.8	31'	91%	144° E	+G	
15th	Lib	16h 1m	-16° 47'	-12.1	31'	96%	157° E	FM	Antares
16th	Oph	16h 55m	-19° 45'	-12.5	31'	99%	169° E	FM	Jupiter, Antares
17th	Sgr	17h 50m	-21° 38'	-12.7	31'	100%	178° W	FM	Jupiter
18th	Sgr	18h 45m	-22° 22'	-12.4	30'	99%	165° W	FM	Saturn
19th	Sgr	19h 39m	-21° 58'	-12.1	30'	95%	153° W	FM	Saturn
20th	Cap	20h 31m	-20° 32'	-11.8	30'	90%	141° W	-G	

Mercury and Venus

Mercury
15th

Venus
15th

Mercury

Date	Con.	R.A.	Dec.	Mag.	Diam.	Ill.	Elon.	Vis.	Rat.	Close To
11th	Gem	6h 50m	24° 58'	-0.3	6"	63%	23° E	PM	***	Mars
13th	Gem	7h 3m	24° 33'	-0.1	7"	59%	24° E	PM	***	Mars
15th	Gem	7h 16m	24° 3'	0.0	7"	55%	25° E	PM	***	Mars
17th	Gem	7h 27m	23° 28'	0.1	7"	51%	26° E	PM	****	Mars
19th	Gem	7h 38m	22° 50'	0.3	7"	47%	27° E	PM	****	Mars

Venus

Date	Con.	R.A.	Dec.	Mag.	Diam.	Ill.	Elon.	Vis.	Rat.	Close To
11th	Tau	4h 4m	19° 49'	-3.9	10"	95%	18° W	AM	*	Pleiades, Hyades, Aldebaran
13th	Tau	4h 14m	20° 21'	-3.9	10"	96%	18° W	AM	*	Pleiades, Hyades, Aldebaran
15th	Tau	4h 24m	20° 50'	-3.9	10"	96%	17° W	AM	*	Pleiades, Hyades, Aldebaran
17th	Tau	4h 35m	21° 16'	-3.9	10"	96%	17° W	AM	*	Hyades, Aldebaran
19th	Tau	4h 45m	21° 41'	-3.9	10"	96%	16° W	AM	*	Hyades, Aldebaran

Mars and the Outer Planets

Mars
15th

Jupiter
15th

Saturn
15th

Mars

Date	Con.	R.A.	Dec.	Mag.	Diam.	Ill.	Elon.	Vis.	Rat.	Close To
11th	Gem	7h 14m	23° 31'	1.8	4"	98%	29° E	PM	*	Mercury
15th	Gem	7h 25m	23° 11'	1.8	4"	98%	28° E	PM	*	Mercury
20th	Gem	7h 38m	22° 41'	1.8	4"	98%	26° E	PM	*	Mercury

The Outer Planets

Planet	Date	Con.	R.A.	Dec.	Mag.	Diam.	Elon.	Vis.	Rat.	Close To
Jupiter	15th	Oph	17h 11m	-22° 24'	-2.6	46"	174° E	AN	*****	
Saturn	15th	Sgr	19h 21m	-21° 45'	0.2	18"	153° W	AM	****	
Uranus	15th	Ari	2h 12m	12° 50'	5.9	3"	50° W	AM	*	
Neptune	15th	Aqr	23h 20m	-5° 24'	7.9	2"	94° W	AM	***	

Highlights

Date	Time (UT)	Event
12th	11:21	The waxing gibbous Moon is north of the bright star Spica. (Virgo, evening sky.)
16th	01:11	The nearly full Moon is north of the bright star Antares. (Scorpius, evening sky.)
	17:34	The nearly full Moon is north of Jupiter. (Evening sky.)
17th	08:31	Full Moon. (Visible all night.)
	15:34	Venus is 4.8° north of the bright star Aldebaran. (Taurus, morning sky.)
18th	14:41	Mercury is 0.2° north of Mars. (Evening sky.)
19th	04:42	The waning gibbous Moon is south of Saturn. (Morning sky.)
	11:27	The waning gibbous Moon is south of dwarf planet Pluto. (Morning sky.)

Planet Locations – June 15th

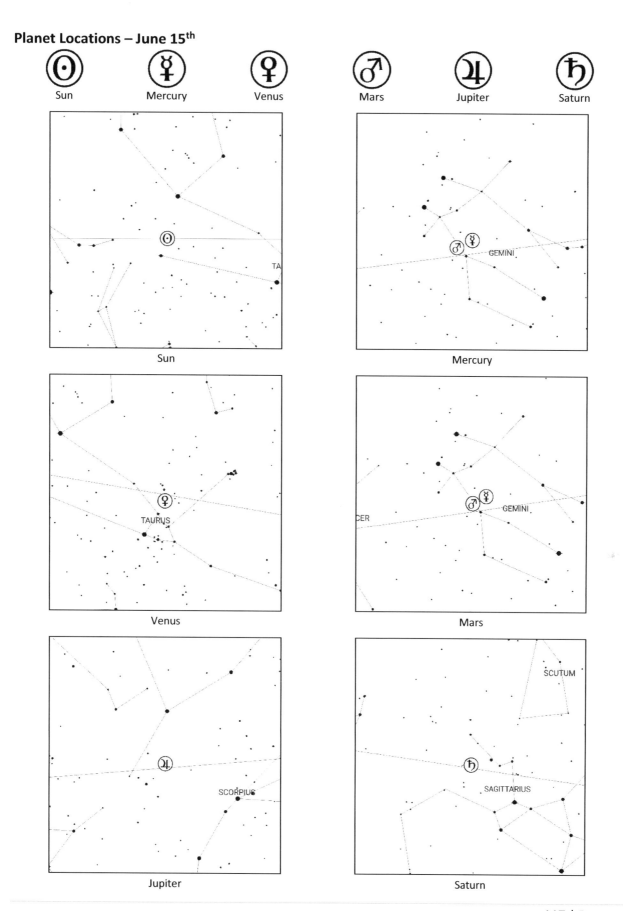

Sun · Mercury · Venus · Mars · Jupiter · Saturn

Sun

Mercury

Venus

Mars

Jupiter

Saturn

June 21st – 30th

The Moon

| 21st | 23rd | 25th | 27th | 29th |

Date	Con	R.A.	Dec	Mag	Diam	Ill.	Elon.	Phase	Close To
21st	Cap	21h 21m	-18° 11'	-11.5	30'	84%	129° W	-G	
22nd	Aqr	22h 9m	-15° 5'	-11.2	30'	76%	118° W	-G	
23rd	Aqr	22h 55m	-11° 24'	-10.8	30'	68%	108° W	-G	Neptune
24th	Aqr	23h 40m	-7° 17'	-10.5	30'	59%	98° W	LQ	Neptune
25th	Psc	0h 25m	-2° 51'	-10.1	30'	49%	88° W	LQ	
26th	Cet	1h 10m	1° 44'	-9.6	30'	39%	78° W	LQ	
27th	Psc	1h 55m	6° 21'	-9.1	30'	30%	67° W	-Cr	Uranus
28th	Ari	2h 43m	10° 49'	-8.5	31'	21%	56° W	-Cr	Uranus
29th	Tau	3h 34m	14° 55'	-7.7	31'	13%	45° W	-Cr	Pleiades
30th	Tau	4h 28m	18° 23'	-6.8	32'	7%	32° W	NM	Pleiades, Hyades, Aldebaran

Mercury and Venus

Mercury
25th

Venus
25th

Mercury

Date	Con.	R.A.	Dec.	Mag.	Diam.	Ill.	Elon.	Vis.	Rat.	Close To
21st	Gem	7h 47m	22° 8'	0.4	8"	44%	27° E	PM	****	Mars
23rd	Gem	7h 56m	21° 25'	0.5	8"	40%	27° E	PM	****	Mars
25th	Cnc	8h 3m	20° 41'	0.7	8"	37%	27° E	PM	****	Mars, Praesepe
27th	Cnc	8h 10m	19° 57'	0.8	9"	33%	26° E	PM	****	Mars, Praesepe
29th	Cnc	8h 15m	19° 13'	1.0	9"	29%	26° E	PM	****	Mars, Praesepe

Venus

Date	Con.	R.A.	Dec.	Mag.	Diam.	Ill.	Elon.	Vis.	Rat.	Close To
21st	Tau	4h 55m	22° 3'	-3.9	10"	97%	16° W	AM	*	Hyades, Aldebaran
23rd	Tau	5h 6m	22° 22'	-3.9	10"	97%	15° W	AM	*	Hyades, Aldebaran
25th	Tau	5h 16m	22° 39'	-3.9	10"	97%	15° W	NV	N/A	
27th	Tau	5h 27m	22° 54'	-3.9	10"	97%	14° W	NV	N/A	
29th	Tau	5h 37m	23° 6'	-3.9	10"	98%	14° W	NV	N/A	

Mars and the Outer Planets

Mars
25th

Jupiter
25th

Saturn
25th

Mars

Date	Con.	R.A.	Dec.	Mag.	Diam.	Ill.	Elon.	Vis.	Rat.	Close To
21st	Gem	7h 41m	22° 35'	1.8	4"	98%	26° E	PM	*	Mercury
25th	Gem	7h 52m	22° 8'	1.8	4"	99%	24° E	PM	*	Mercury
30th	Cnc	8h 5m	21° 30'	1.8	4"	99%	22° E	PM	*	Mercury, Praesepe

The Outer Planets

Planet	Date	Con.	R.A.	Dec.	Mag.	Diam.	Elon.	Vis.	Rat.	Close To
Jupiter	25th	Oph	17h 6m	-22° 19'	-2.6	46"	163° E	PM	*****	Antares
Saturn	25th	Sgr	19h 18m	-21° 52'	0.1	18"	164° W	AM	****	
Uranus	25th	Ari	2h 14m	12° 58'	5.8	3"	60° W	AM	*	
Neptune	25th	Aqr	23h 20m	-5° 24'	7.9	2"	104° W	AM	***	

Highlights

Date	Time (UT)	Event
21st	15:55	Summer Solstice.
	23:46	Neptune is stationary prior to beginning retrograde motion. (Morning sky.)
23rd	23:03	Mercury is at greatest eastern elongation from the Sun. (Evening sky.)
	23:30	The waning gibbous Moon is south of Neptune. (Morning sky.)
25th	09:47	Last Quarter Moon. (Morning sky.)
27th	21:06	The waning crescent Moon is south of Uranus. (Morning sky.)
	N/A	The Boötid meteor shower is at its maximum. (ZHR: Variable.)
29th	18:32	The waning crescent Moon is south of the Pleiades star cluster. (Taurus, morning sky.)
	N/A	Good opportunity to see Earthshine on the waning crescent Moon. (Morning sky.)
30th	16:21	The waning crescent Moon is north of the bright star Aldebaran. (Taurus, morning sky.)

Planet Locations – June 25th

| Sun ☉ | Mercury ☿ | Venus ♀ | Mars ♂ | Jupiter ♃ | Saturn ♄ |

Sun

Mercury

Venus

Mars

Jupiter

Saturn

The Moon

1st   3rd   5th   7th   9th

Date	Con	R.A.	Dec	Mag	Diam	Ill.	Elon.	Phase	Close To
1st	Tau	5h 26m	20° 56'	-5.7	32'	2%	19° W	NM	Venus
2nd	Gem	6h 27m	22° 15'	-4.3	32'	0%	4° W	NM	Venus
3rd	Gem	7h 29m	22° 7'	-4.9	33'	1%	10° E	NM	Mars
4th	Cnc	8h 32m	20° 30'	-6.2	33'	4%	25° E	NM	Mercury, Mars, Praesepe
5th	Leo	9h 33m	17° 31'	-7.3	33'	10%	39° E	NM	Regulus
6th	Leo	10h 31m	13° 25'	-8.2	33'	18%	53° E	+Cr	Regulus
7th	Leo	11h 27m	8° 34'	-9.0	33'	28%	65° E	+Cr	
8th	Vir	12h 20m	3° 19'	-9.6	32'	39%	78° E	FQ	
9th	Vir	13h 12m	-2° 1'	-10.1	32'	51%	90° E	FQ	Spica
10th	Vir	14h 4m	-7° 9'	-10.6	32'	62%	102° E	FQ	Spica

Mercury and Venus

Mercury
5th

Venus
5th

Mercury

Date	Con.	R.A.	Dec.	Mag.	Diam.	Ill.	Elon.	Vis.	Rat.	Close To
1st	Cnc	8h 19m	18° 31'	1.2	9"	26%	25° E	PM	****	Mars, Praesepe
3rd	Cnc	8h 22m	17° 51'	1.4	10"	23%	23° E	PM	***	Mars, Praesepe
5th	Cnc	8h 24m	17° 14'	1.6	10"	19%	22° E	PM	***	Mars, Praesepe
7th	Cnc	8h 24m	16° 41'	1.9	11"	16%	20° E	PM	***	Mars, Praesepe
9th	Cnc	8h 23m	16° 13'	2.2	11"	12%	18° E	PM	***	Mars, Praesepe

Venus

Date	Con.	R.A.	Dec.	Mag.	Diam.	Ill.	Elon.	Vis.	Rat.	Close To
1st	Tau	5h 48m	23° 15'	-3.9	10"	98%	13° W	NV	N/A	Moon
3rd	Tau	5h 59m	23° 21'	-3.9	10"	98%	12° W	NV	N/A	
5th	Gem	6h 9m	23° 25'	-3.9	10"	98%	12° W	NV	N/A	
7th	Gem	6h 20m	23° 26'	-3.9	10"	98%	11° W	NV	N/A	
9th	Gem	6h 31m	23° 24'	-3.9	10"	99%	11° W	NV	N/A	

Mars and the Outer Planets

Mars
5th

Jupiter
5th

Saturn
5th

Mars

Date	Con.	R.A.	Dec.	Mag.	Diam.	Ill.	Elon.	Vis.	Rat.	Close To
1st	Cnc	8h 8m	21° 22'	1.8	4"	99%	22° E	PM	*	Mercury, Praesepe
5th	Cnc	8h 19m	20° 48'	1.8	4"	99%	20° E	PM	*	Mercury, Praesepe
10th	Cnc	8h 32m	20° 3'	1.8	4"	99%	19° E	PM	*	Mercury, Praesepe

The Outer Planets

Planet	Date	Con.	R.A.	Dec.	Mag.	Diam.	Elon.	Vis.	Rat.	Close To
Jupiter	5th	Oph	17h 1m	-22° 14'	-2.6	45"	151° E	PM	****	Antares
Saturn	5th	Sgr	19h 15m	-21° 58'	0.1	18"	175° W	AN	*****	
Uranus	5th	Ari	2h 15m	13° 4'	5.8	3"	70° W	AM	**	
Neptune	5th	Aqr	23h 19m	-5° 26'	7.9	2"	114° W	AM	****	

Highlights

Date	Time (UT)	Event
2nd	19:17	New Moon. (Not visible.)
	19:23	Total solar eclipse. Visible from the southern Pacific and southern South America.
4th	04:45	The waxing crescent Moon is south of Mars. (Evening sky.)
	07:23	The waxing crescent Moon is north of Mercury. (Evening sky.)
	15:29	The waxing crescent Moon is north of the Praesepe star cluster. (Cancer, evening sky.)
5th	N/A	Good opportunity to see Earthshine on the waxing crescent Moon. (Evening sky.)
6th	02:30	The waxing crescent Moon is north of the bright star Regulus. (Leo, evening sky.)
7th	04:16	Mercury is stationary prior to beginning retrograde motion. (Evening sky.)
	13:14	Mercury is 3.8° south of Mars. (Evening sky.)
9th	10:55	First Quarter Moon. (Evening sky.)
	16:08	Saturn is at opposition. (Visible all night.)
	17:22	The first quarter Moon is north of the bright star Spica. (Virgo, evening sky.)

Planet Locations – July 5th

| ☉ Sun | ☿ Mercury | ♀ Venus | ♂ Mars | ♃ Jupiter | ♄ Saturn |

Sun

Mercury

Venus

Mars

Jupiter

Saturn

July 11th – 20th

The Moon

| | 11th | | 13th | | 15th | | 17th | | 19th |

Date	Con	R.A.	Dec	Mag	Diam	Ill.	Elon.	Phase	Close To
11th	Lib	14h 55m	-11° 50'	-11.0	31'	72%	113° E	+G	
12th	Lib	15h 47m	-15° 52'	-11.4	31'	81%	125° E	+G	Antares
13th	Oph	16h 41m	-19° 2'	-11.7	31'	89%	138° E	+G	Jupiter, Antares
14th	Oph	17h 34m	-21° 12'	-12.0	31'	94%	150° E	+G	Jupiter
15th	Sgr	18h 29m	-22° 16'	-12.3	30'	98%	163° E	FM	Saturn
16th	Sgr	19h 22m	-22° 12'	-12.6	30'	100%	175° E	FM	Saturn
17th	Cap	20h 15m	-21° 5'	-12.6	30'	100%	173° W	FM	
18th	Cap	21h 6m	-18° 60'	-12.3	30'	98%	161° W	FM	
19th	Cap	21h 55m	-16° 6'	-12.0	30'	94%	150° W	-G	
20th	Aqr	22h 41m	-12° 35'	-11.7	29'	88%	139° W	-G	Neptune

Mercury and Venus

Mercury
15th

Venus
15th

Mercury

Date	Con.	R.A.	Dec.	Mag.	Diam.	Ill.	Elon.	Vis.	Rat.	Close To
11th	Cnc	8h 21m	15° 50'	2.6	11"	9%	15° E	NV	N/A	Mars, Praesepe
13th	Cnc	8h 18m	15° 34'	3.1	11"	7%	12° E	NV	N/A	Mars, Praesepe
15th	Cnc	8h 14m	15° 24'	3.6	12"	4%	9° E	NV	N/A	Mars, Praesepe
17th	Cnc	8h 9m	15° 21'	4.1	12"	3%	6° E	NV	N/A	Praesepe
19th	Cnc	8h 3m	15° 26'	4.6	12"	1%	2° E	NV	N/A	Venus, Praesepe

Venus

Date	Con.	R.A.	Dec.	Mag.	Diam.	Ill.	Elon.	Vis.	Rat.	Close To
11th	Gem	6h 41m	23° 20'	-3.9	10"	99%	10° W	NV	N/A	
13th	Gem	6h 52m	23° 12'	-3.9	10"	99%	9° W	NV	N/A	
15th	Gem	7h 3m	23° 2'	-3.9	10"	99%	9° W	NV	N/A	
17th	Gem	7h 13m	22° 49'	-3.9	10"	99%	8° W	NV	N/A	
19th	Gem	7h 24m	22° 34'	-3.9	10"	99%	7° W	NV	N/A	Mercury

Mars and the Outer Planets

Mars	Jupiter	Saturn
15th	15th	15th

Mars

Date	Con.	R.A.	Dec.	Mag.	Diam.	Ill.	Elon.	Vis.	Rat.	Close To
11th	Cnc	8h 34m	19° 54'	1.8	4"	99%	18° E	PM	*	Mercury, Praesepe
15th	Cnc	8h 45m	19° 15'	1.8	4"	99%	17° E	PM	*	Mercury, Praesepe
20th	Cnc	8h 58m	18° 23'	1.8	4"	99%	15° E	PM	*	Praesepe

The Outer Planets

Planet	Date	Con.	R.A.	Dec.	Mag.	Diam.	Elon.	Vis.	Rat.	Close To
Jupiter	15th	Oph	16h 57m	-22° 10'	-2.5	44"	140° E	PM	****	Antares
Saturn	15th	Sgr	19h 12m	-22° 4'	0.1	18"	174° E	AN	*****	Moon
Uranus	15th	Ari	2h 16m	13° 9'	5.8	4"	80° W	AM	**	
Neptune	15th	Aqr	23h 19m	-5° 28'	7.8	2"	125° W	AM	****	

Highlights

Date	Time (UT)	Event
13th	07:19	The waxing gibbous Moon is north of the bright star Antares. (Scorpius, evening sky.)
	18:54	The waxing gibbous Moon is north of Jupiter. (Evening sky.)
14th	10:33	Dwarf planet Pluto is at opposition. (Visible all night.)
16th	08:13	The almost full Moon is south of Saturn. (Visible all night.)
	15:41	The almost full Moon is south of dwarf planet Pluto. (Visible all night.)
	21:30	Partial lunar eclipse. Visible from Africa, Antarctica, Asia, the Atlantic, Australia, Europe and South America.
	21:39	Full Moon. (Visible all night.)
19th	04;32	Dwarf planet Ceres is stationary prior to resuming prograde motion. (Evening sky.)

Planet Locations – July 15th

☉ Sun	☿ Mercury	♀ Venus	♂ Mars	♃ Jupiter	♄ Saturn

Sun

Mercury

Venus

Mars

Jupiter

Saturn

The Moon

| 21st | 23rd | 25th | 27th | 29th | 31st |

Date	Con	R.A.	Dec	Mag	Diam	Ill.	Elon.	Phase	Close To
21st	Aqr	23h 27m	-8° 35'	-11.4	29'	82%	129° W	-G	Neptune
22nd	Psc	0h 11m	-4° 15'	-11.1	30'	74%	119° W	-G	
23rd	Cet	0h 55m	0° 16'	-10.7	30'	65%	109° W	LQ	
24th	Psc	1h 40m	4° 49'	-10.4	30'	55%	98° W	LQ	Uranus
25th	Cet	2h 26m	9° 17'	-9.9	30'	46%	88° W	LQ	Uranus
26th	Ari	3h 15m	13° 27'	-9.4	31'	36%	77° W	-Cr	Pleiades
27th	Tau	4h 6m	17° 8'	-8.8	31'	26%	65° W	-Cr	Pleiades, Hyades, Aldebaran
28th	Tau	5h 2m	20° 2'	-8.1	32'	17%	52° W	-Cr	Hyades, Aldebaran
29th	Ori	6h 1m	21° 52'	-7.3	32'	9%	38° W	NM	
30th	Gem	7h 3m	22° 22'	-6.2	33'	4%	24° W	NM	Mercury
31st	Cnc	8h 6m	21° 21'	-4.9	33'	1%	9° W	NM	Mercury, Venus, Praesepe

Mercury and Venus

Mercury
25th

Venus
25th

Mercury

Date	Con.	R.A.	Dec.	Mag.	Diam.	Ill.	Elon.	Vis.	Rat.	Close To
21st	Cnc	7h 58m	15° 36'	4.8	12"	1%	1° W	NV	N/A	Venus
23rd	Gem	7h 52m	15° 52'	4.5	11"	1%	4° W	NV	N/A	Venus
25th	Gem	7h 48m	16° 13'	4.0	11"	3%	8° W	NV	N/A	Venus
27th	Gem	7h 44m	16° 37'	3.3	11"	5%	10° W	NV	N/A	Venus
29th	Gem	7h 41m	17° 4'	2.7	10"	8%	13° W	NV	N/A	Venus
31st	Gem	7h 40m	17° 31'	2.1	10"	12%	15° W	AM	**	Moon

Venus

Date	Con.	R.A.	Dec.	Mag.	Diam.	Ill.	Elon.	Vis.	Rat.	Close To
21st	Gem	7h 35m	22° 16'	-3.9	10"	99%	7° W	NV	N/A	Mercury
23rd	Gem	7h 45m	21° 55'	-3.9	10"	99%	6° W	NV	N/A	Mercury
25th	Gem	7h 56m	21° 32'	-3.9	10"	100%	6° W	NV	N/A	Mercury
27th	Cnc	8h 6m	21° 6'	-3.9	10"	100%	5° W	NV	N/A	Mercury, Praesepe
29th	Cnc	8h 16m	20° 38'	-3.9	10"	100%	4° W	NV	N/A	Mercury, Praesepe
31st	Cnc	8h 27m	20° 7'	-3.9	10"	100%	4° W	NV	N/A	Moon, Praesepe

Mars and the Outer Planets

Mars
25th

Jupiter
25th

Saturn
25th

Mars

Date	Con.	R.A.	Dec.	Mag.	Diam.	Ill.	Elon.	Vis.	Rat.	Close To
21st	Cnc	9h 0m	18° 12'	1.8	4"	99%	15° E	NV	N/A	Praesepe
25th	Cnc	9h 11m	17° 28'	1.8	4"	100%	13° E	NV	N/A	Praesepe
31st	Leo	9h 26m	16° 18'	1.8	4"	100%	11° E	NV	N/A	

The Outer Planets

Planet	Date	Con.	R.A.	Dec.	Mag.	Diam.	Elon.	Vis.	Rat.	Close To
Jupiter	25th	Oph	16h 54m	-22° 8'	-2.5	43"	129° E	PM	****	Antares
Saturn	25th	Sgr	19h 9m	-22° 11'	0.1	18"	163° E	PM	****	
Uranus	25th	Ari	2h 17m	13° 12'	5.8	4"	90° W	AM	**	Moon
Neptune	25th	Aqr	23h 19m	-5° 32'	7.8	2"	135° W	AM	****	

Highlights

Date	Time (UT)	Event
21st	09:07	The waning gibbous Moon is south of Neptune. (Morning sky.)
	12:28	Mercury is at inferior conjunction with the Sun. (Not visible.)
25th	01:19	Last Quarter Moon. (Morning sky.)
	07:47	The last quarter Moon is south of Uranus. (Morning sky.)
27th	01:38	The waning crescent Moon is south of the Pleaides star cluster. (Taurus, morning sky.)
	23:46	The waning crescent Moon is north of the bright star Aldebaran. (Taurus, morning sky.)
28th	N/A	The Piscis Austrinid meteor shower is at its maximum. (ZHR: 5)
	N/A	Good opportunity to see Earthshine on the waning crescent Moon. (Morning sky.)
30th	N/A	The Delta Aquariid meteor shower is at its maximum. (ZHR: 16)
	N/A	The Alpha Capricornid meteor shower is at its maximum. (ZHR: 5)
31st	18:47	Mercury is stationary prior to resuming prograde motion. (Morning sky.)

Planet Locations – July 25th

| Sun | Mercury | Venus | | Mars | Jupiter | Saturn |

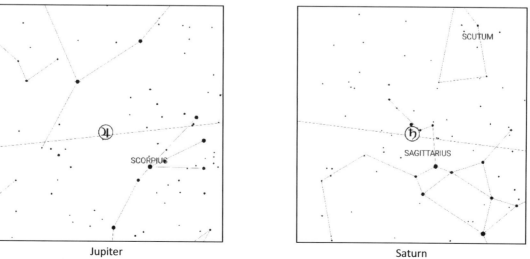

Sun

Mercury

Venus

Mars

Jupiter

Saturn

August 1ˢᵗ to 10ᵗʰ

The Moon

| | 1ˢᵗ | | 3ʳᵈ | | 5ᵗʰ | | 7ᵗʰ | | 9ᵗʰ |

Date	Con	R.A.	Dec	Mag	Diam	Ill.	Elon.	Phase	Close To
1st	Cnc	9h 9m	18° 50'	-4.5	33'	0%	6° E	NM	Venus, Mars, Praesepe
2nd	Leo	10h 10m	15° 2'	-5.9	33'	3%	20° E	NM	Mars, Regulus
3rd	Leo	11h 8m	10° 15'	-7.1	33'	8%	34° E	NM	
4th	Vir	12h 4m	4° 55'	-8.0	33'	16%	47° E	+Cr	
5th	Vir	12h 58m	0° 35'	-8.8	33'	26%	59° E	+Cr	Spica
6th	Vir	13h 51m	-5° 56'	-9.5	32'	36%	72° E	+Cr	Spica
7th	Lib	14h 43m	-10° 50'	-10.0	32'	47%	84° E	FQ	
8th	Lib	15h 35m	-15° 4'	-10.5	31'	58%	96° E	FQ	
9th	Oph	16h 28m	-18° 27'	-10.9	31'	68%	108° E	+G	Jupiter, Antares
10th	Oph	17h 22m	-20° 50'	-11.2	31'	78%	121° E	+G	Jupiter

Mercury and Venus

Mercury
5ᵗʰ

Venus
5ᵗʰ

Mercury

Date	Con.	R.A.	Dec.	Mag.	Diam.	Ill.	Elon.	Vis.	Rat.	Close To
1st	Gem	7h 40m	17° 45'	1.9	10"	14%	16° W	AM	**	
3rd	Gem	7h 42m	18° 11'	1.4	9"	19%	18° W	AM	***	
5th	Gem	7h 45m	18° 34'	0.9	9"	25%	19° W	AM	***	
7th	Gem	7h 50m	18° 54'	0.5	8"	32%	20° W	AM	***	
9th	Cnc	7h 57m	19° 8'	0.2	8"	38%	20° W	AM	***	

Venus

Date	Con.	R.A.	Dec.	Mag.	Diam.	Ill.	Elon.	Vis.	Rat.	Close To
1st	Cnc	8h 32m	19° 51'	-3.9	10"	100%	3° W	NV	N/A	Moon, Praesepe
3rd	Cnc	8h 42m	19° 17'	-3.9	10"	100%	3° W	NV	N/A	Praesepe
5th	Cnc	8h 52m	18° 41'	-3.9	10"	100%	2° W	NV	N/A	Praesepe
7th	Cnc	9h 2m	18° 2'	-3.9	10"	100%	2° W	NV	N/A	Praesepe
9th	Cnc	9h 12m	17° 22'	-3.9	10"	100%	1° W	NV	N/A	Mars, Praesepe

Mars and the Outer Planets

Mars
5th

Jupiter
5th

Saturn
5th

Mars

Date	Con.	R.A.	Dec.	Mag.	Diam.	Ill.	Elon.	Vis.	Rat.	Close To
1st	Leo	9h 28m	16° 6'	1.8	4"	100%	11° E	NV	N/A	Moon, Regulus
5th	Leo	9h 38m	15° 17'	1.8	4"	100%	9° E	NV	N/A	Regulus
10th	Leo	9h 51m	14° 14'	1.8	4"	100%	8° E	NV	N/A	Venus, Regulus

The Outer Planets

Planet	Date	Con.	R.A.	Dec.	Mag.	Diam.	Elon.	Vis.	Rat.	Close To
Jupiter	5th	Oph	16h 53m	-22° 7'	-2.4	42"	118° E	PM	****	Antares
Saturn	5th	Sgr	19h 6m	-22° 17'	0.2	18"	151° E	PM	****	
Uranus	5th	Ari	2h 17m	13° 14'	5.8	4"	101° W	AM	**	
Neptune	5th	Aqr	23h 18m	-5° 37'	7.8	2"	146° W	AM	****	

Highlights

Date	Time (UT)	Event
1st	03:12	New Moon. (Not visible.)
4th	N/A	Good opportunity to see Earthshine in the waxing crescent Moon. (Evening sky.)
6th	01:06	The nearly first quarter Moon is north of the bright star Spica. (Evening sky.)
7th	17:32	First Quarter Moon. (Evening sky.)
9th	11:10	The waxing gibbous Moon is north of the bright star Antares. (Evening sky.)
	22:55	Mercury is at greatest western elongation from the Sun. (Morning sky.)
10th	00:00	The waxing gibbous Moon is north of Jupiter. (Evening sky.)

Planet Locations – August 5th

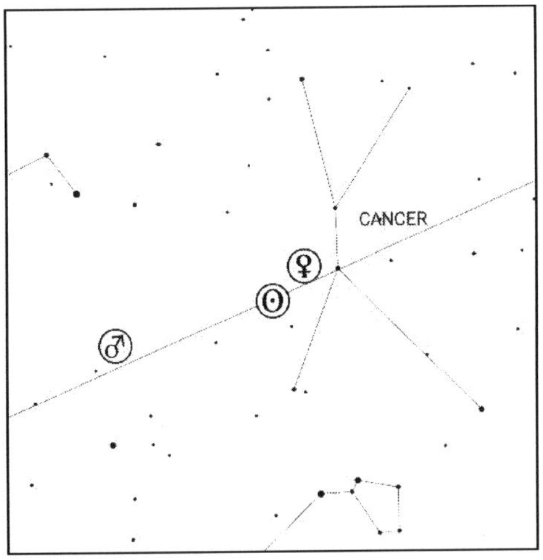

Sun 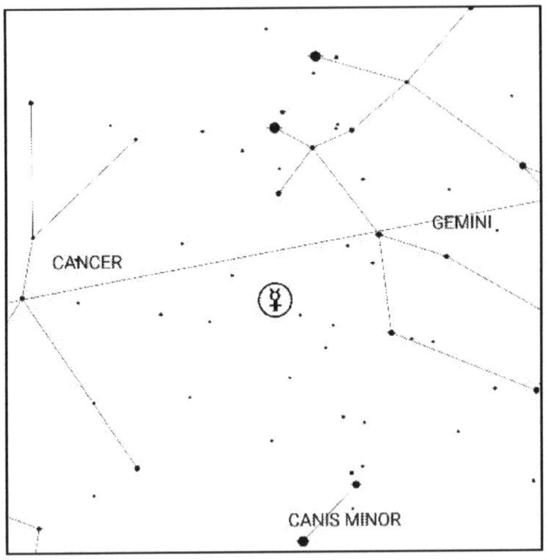   Mercury 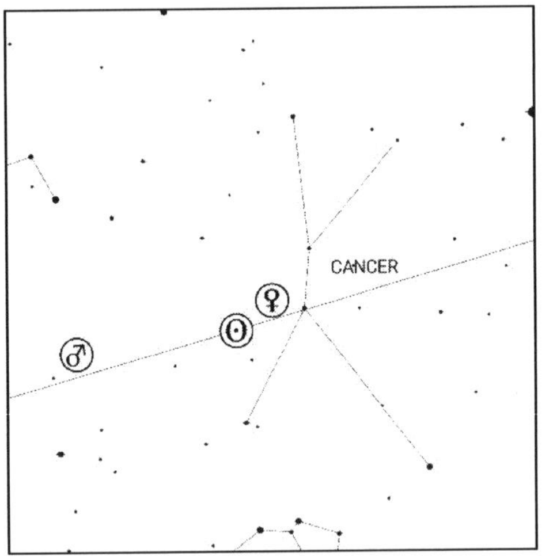   Venus 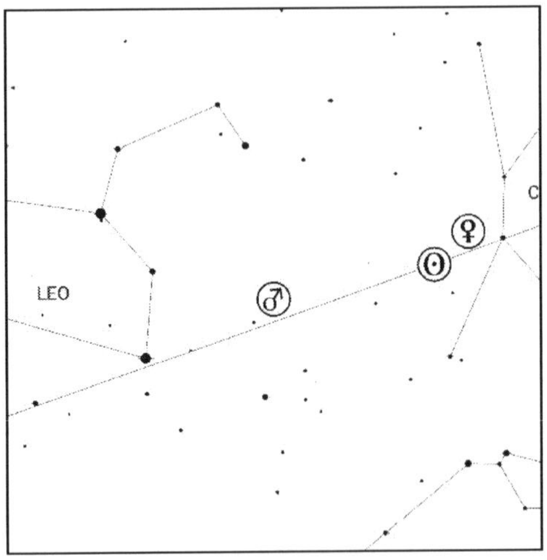   Mars 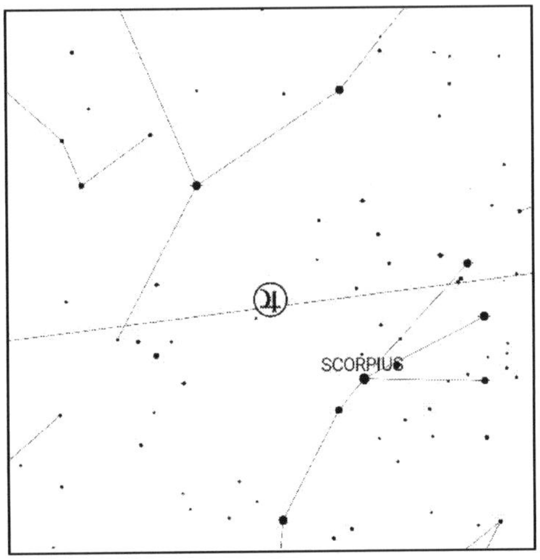   Jupiter   Saturn

Sun

Mercury

Venus

Mars

Jupiter

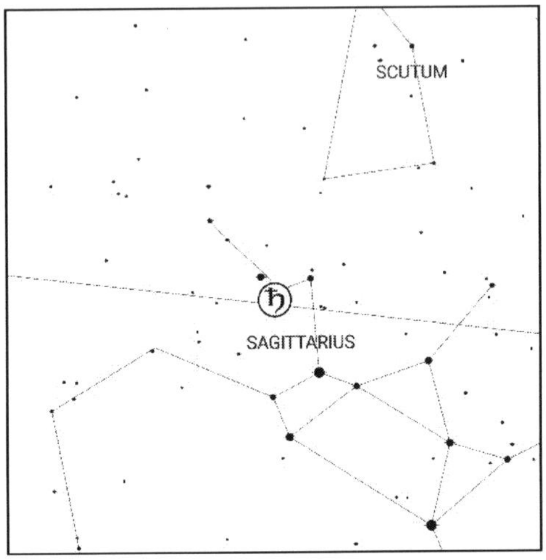

Saturn

The Moon

| | 11th | | 13th | | 15th | | 17th | | 19th |

Date	Con	R.A.	Dec	Mag	Diam	Ill.	Elon.	Phase		Close To
11th	Sgr	18h 16m	-22° 9'	-11.5	30'	85%	133° E	+G		
12th	Sgr	19h 9m	-22° 21'	-11.9	30'	92%	145° E	+G		Saturn
13th	Sgr	20h 2m	-21° 29'	-12.1	30'	96%	158° E	FM		
14th	Cap	20h 52m	-19° 37'	-12.4	30'	99%	169° E	FM		
15th	Cap	21h 42m	-16° 56'	-12.7	30'	100%	179° W	FM		
16th	Aqr	22h 29m	-13° 32'	-12.4	29'	99%	168° W	FM		
17th	Aqr	23h 15m	-9° 38'	-12.2	29'	96%	158° W	FM		Neptune
18th	Psc	23h 59m	-5° 22'	-11.9	29'	92%	148° W	-G		Neptune
19th	Cet	0h 43m	0° 53'	-11.6	30'	86%	138° W	-G		
20th	Psc	1h 27m	3° 39'	-11.3	30'	79%	127° W	-G		

Mercury and Venus

Mercury
15th

Venus
15th

Mercury

Date	Con.	R.A.	Dec.	Mag.	Diam.	Ill.	Elon.	Vis.	Rat.	Close To
11th	Cnc	8h 6m	19° 15'	-0.1	7"	46%	19° W	AM	***	Praesepe
13th	Cnc	8h 16m	19° 13'	-0.4	7"	54%	19° W	AM	***	Praesepe
15th	Cnc	8h 28m	19° 3'	-0.6	6"	61%	18° W	AM	**	Praesepe
17th	Cnc	8h 41m	18° 41'	-0.8	6"	69%	16° W	AM	**	Praesepe
19th	Cnc	8h 55m	18° 9'	-1.0	6"	76%	15° W	NV	N/A	Praesepe

Venus

Date	Con.	R.A.	Dec.	Mag.	Diam.	Ill.	Elon.	Vis.	Rat.	Close To
11th	Cnc	9h 22m	16° 40'	-3.9	10"	100%	0° W	NV	N/A	Mars
13th	Leo	9h 32m	15° 55'	-3.9	10"	100%	0° E	NV	N/A	Mars, Regulus
15th	Leo	9h 42m	15° 9'	-3.9	10"	100%	1° E	NV	N/A	Mars, Regulus
17th	Leo	9h 51m	14° 22'	-3.9	10"	100%	1° E	NV	N/A	Mars, Regulus
19th	Leo	10h 1m	13° 33'	-3.9	10"	100%	2° E	NV	N/A	Mars, Regulus

Mars and the Outer Planets

Mars
15th

Jupiter
15th

Saturn
15th

Mars

Date	Con.	R.A.	Dec.	Mag.	Diam.	Ill.	Elon.	Vis.	Rat.	Close To
11th	Leo	9h 53m	14° 1'	1.8	4"	100%	7° E	NV	N/A	Venus, Regulus
15th	Leo	10h 3m	13° 8'	1.8	4"	100%	6° E	NV	N/A	Venus, Regulus
20th	Leo	10h 15m	12° 0'	1.8	4"	100%	4° E	NV	N/A	Venus, Regulus

The Outer Planets

Planet	Date	Con.	R.A.	Dec.	Mag.	Diam.	Elon.	Vis.	Rat.	Close To
Jupiter	15th	Oph	16h 53m	-22° 9'	-2.3	41"	109° E	PM	***	Antares
Saturn	15th	Sgr	19h 4m	-22° 21'	0.2	18"	141° E	PM	****	
Uranus	15th	Ari	2h 17m	13° 15'	5.8	4"	110° W	AM	***	
Neptune	15th	Aqr	23h 17m	-5° 43'	7.8	2"	155° W	AM	*****	

Highlights

Date	Time (UT)	Event
11th	16:03	Jupiter is stationary prior to resuming prograde motion. (Evening sky.)
12th	02:18	Uranus is stationary prior to beginning retrograde motion. (Morning sky.)
	09:50	The waxing gibbous Moon is south of Saturn. (Evening sky.)
	21:30	The waxing gibbous Moon is south of the dwarf planet Pluto. (Evening sky.)
13th	N/A	The Perseid meteor shower is at its maximum. (ZHR: 100)
14th	05:35	Venus is at superior conjunction with the Sun. (Not visible.)
15th	12:30	Full Moon. (Visible all night.)
17th	06:28	Mercury is 0.9° south of the Praesepe star cluster. (Cancer, morning sky.)
	13:19	The waning gibbous Moon is south of Neptune. (Morning sky.)

Planet Locations – August 15th

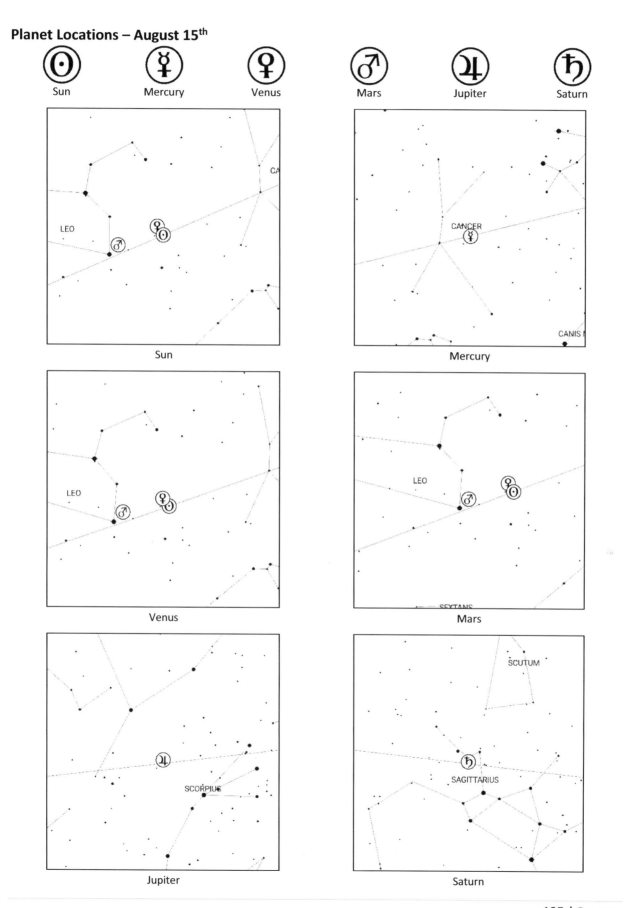

August 21st – 31st

The Moon

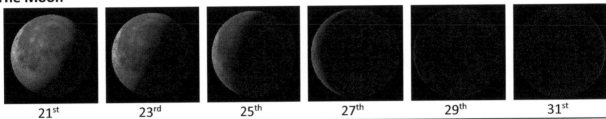

| | 21st | 23rd | 25th | 27th | 29th | 31st |

Date	Con	R.A.	Dec	Mag	Diam	Ill.	Elon.	Phase	Close To
21st	Cet	2h 12m	8° 7'	-10.9	30'	71%	117° W	-G	Uranus
22nd	Ari	2h 59m	12° 20'	-10.6	30'	61%	106° W	LQ	Uranus, Pleiades
23rd	Tau	3h 49m	16° 6'	-10.2	31'	51%	95° W	LQ	Pleiades, Hyades, Aldebaran
24th	Tau	4h 41m	19° 13'	-9.7	31'	41%	83° W	LQ	Hyades, Aldebaran
25th	Tau	5h 37m	21° 25'	-9.1	32'	31%	70° W	-Cr	
26th	Gem	6h 37m	22° 26'	-8.5	32'	21%	56° W	-Cr	
27th	Gem	7h 38m	22° 3'	-7.6	33'	12%	41° W	NM	
28th	Cnc	8h 41m	20° 10'	-6.6	33'	6%	26° W	NM	Praesepe
29th	Leo	9h 43m	16° 52'	-5.4	33'	1%	12° W	NM	Mercury, Regulus
30th	Leo	10h 43m	12° 24'	-4.3	33'	0%	2° E	NM	Mercury, Venus, Mars, Regulus
31st	Vir	11h 41m	7° 6'	-5.6	33'	2%	16° E	NM	Venus

Mercury and Venus

Mercury
25th

Venus
25th

Mercury

Date	Con.	R.A.	Dec.	Mag.	Diam.	Ill.	Elon.	Vis.	Rat.	Close To
21st	Cnc	9h 10m	17° 26'	-1.2	6"	82%	13° W	NV	N/A	Praesepe
23rd	Leo	9h 25m	16° 33'	-1.3	5"	88%	11° W	NV	N/A	
25th	Leo	9h 41m	15° 29'	-1.4	5"	92%	9° W	NV	N/A	Regulus
27th	Leo	9h 56m	14° 18'	-1.5	5"	95%	7° W	NV	N/A	Mars, Regulus
29th	Leo	10h 12m	12° 59'	-1.6	5"	98%	5° W	NV	N/A	Moon, Venus, Mars, Regulus
31st	Leo	10h 27m	11° 35'	-1.7	5"	99%	3° W	NV	N/A	Venus, Mars, Regulus

Venus

Date	Con.	R.A.	Dec.	Mag.	Diam.	Ill.	Elon.	Vis.	Rat.	Close To
21st	Leo	10h 10m	12° 42'	-3.9	10"	100%	2° E	NV	N/A	Mars, Regulus
23rd	Leo	10h 20m	11° 50'	-3.9	10"	100%	3° E	NV	N/A	Mars, Regulus
25th	Leo	10h 29m	10° 57'	-3.9	10"	100%	3° E	NV	N/A	Mars, Regulus
27th	Leo	10h 39m	10° 2'	-3.9	10"	100%	4° E	NV	N/A	Mars, Regulus
29th	Leo	10h 48m	9° 6'	-3.9	10"	100%	4° E	NV	N/A	Mercury, Mars
31st	Leo	10h 57m	8° 10'	-3.9	10"	100%	5° E	NV	N/A	Moon, Mercury, Mars

Mars and the Outer Planets

Mars
25th

Jupiter
25th

Saturn
25th

Mars

Date	Con.	R.A.	Dec.	Mag.	Diam.	Ill.	Elon.	Vis.	Rat.	Close To
21st	Leo	10h 17m	11° 46'	1.8	4"	100%	4° E	NV	N/A	Venus, Regulus
25th	Leo	10h 27m	10° 50'	1.8	4"	100%	3° E	NV	N/A	Venus, Regulus
31st	Leo	10h 41m	9° 25'	1.7	4"	100%	1° E	NV	N/A	Mercury, Venus, Regulus

The Outer Planets

Planet	Date	Con.	R.A.	Dec.	Mag.	Diam.	Elon.	Vis.	Rat.	Close To
Jupiter	25th	Oph	16h 54m	-22° 13'	-2.3	40"	100° E	PM	***	Antares
Saturn	25th	Sgr	19h 2m	-22° 25'	0.3	18"	132° E	PM	****	
Uranus	25th	Ari	2h 17m	13° 13'	5.7	4"	120° W	AM	***	
Neptune	25th	Aqr	23h 16m	-5° 49'	7.8	2"	165° W	AM	*****	

Highlights

Date	Time (UT)	Event
21st	15:01	The waning gibbous Moon is south of Uranus. (Morning sky.)
23rd	12:17	The almost last quarter Moon is south of the Pleiades star cluster. (Taurus, morning sky.)
	14:57	Last Quarter Moon. (Morning sky.)
24th	10:31	The just-past last quarter Moon is north of the bright star Aldebaran. (Taurus, morning sky.)
27th	N/A	Good opportunity to see Earthshine on the waning crescent Moon. (Morning sky.)
28th	11:56	The waning crescent Moon is north of the Praesepe star cluster. (Cancer, morning sky.)
30th	10:38	New Moon. (Not visible.)

Planet Locations – August 25th

Sun Mercury Venus Mars Jupiter Saturn

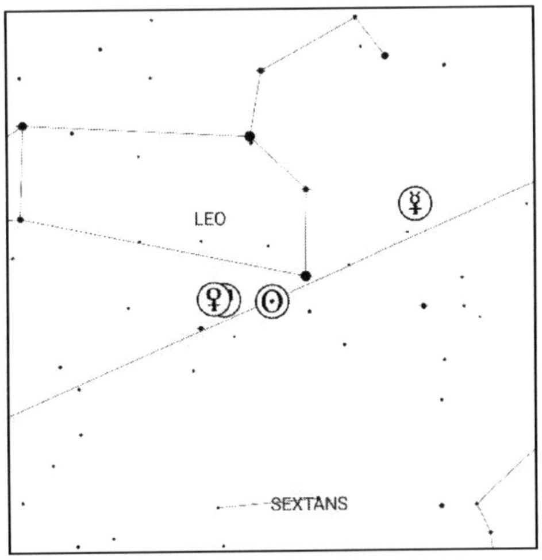

Sun

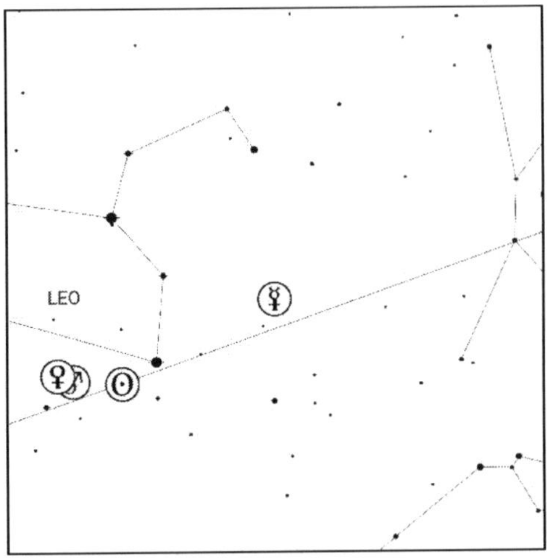

Mercury

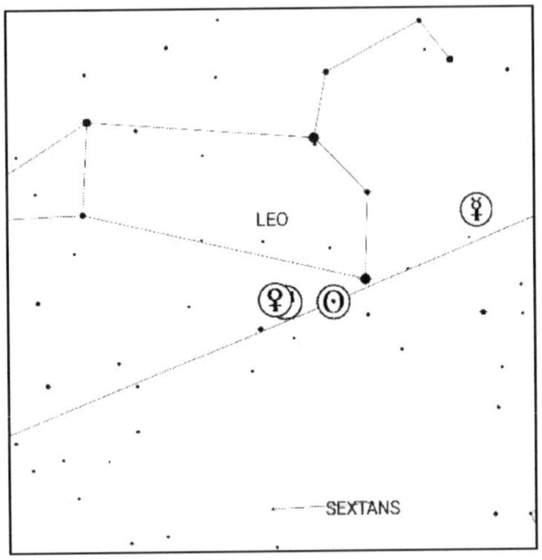

Venus

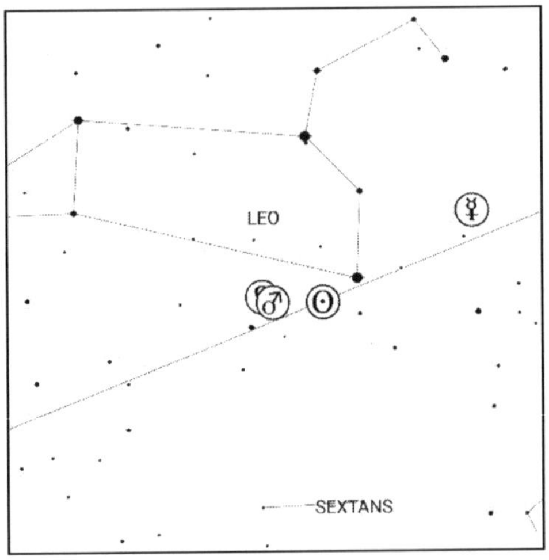

Mars

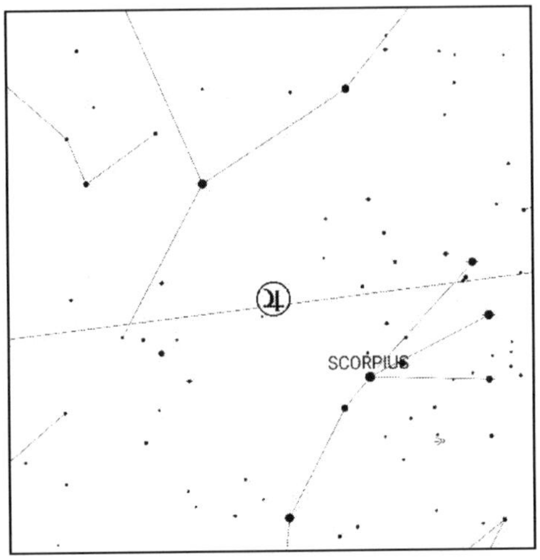

Jupiter

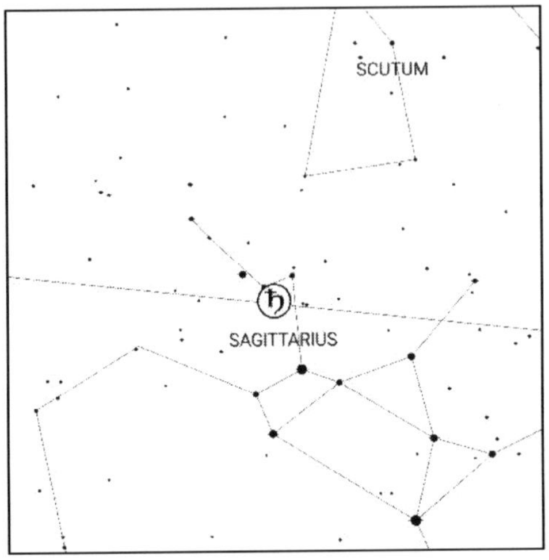

Saturn

The Moon

| 1st | 3rd | 5th | 7th | 9th |

Date	Con	R.A.	Dec	Mag	Diam	Ill.	Elon.	Phase		Close To
1st	Vir	12h 37m	1° 26'	-6.8	33'	7%	29° E	NM		Spica
2nd	Vir	13h 32m	-4° 13'	-7.8	33'	14%	42° E	+Cr		Spica
3rd	Lib	14h 26m	-9° 29'	-8.6	32'	23%	54° E	+Cr		
4th	Lib	15h 20m	-14° 5'	-9.3	32'	33%	67° E	+Cr		
5th	Sco	16h 14m	-17° 49'	-9.8	31'	43%	80° E	FQ		Jupiter, Antares
6th	Oph	17h 8m	-20° 30'	-10.3	31'	54%	92° E	FQ		Jupiter, Antares
7th	Sgr	18h 3m	-22° 5'	-10.7	30'	64%	105° E	FQ		
8th	Sgr	18h 56m	-22° 32'	-11.0	30'	73%	118° E	+G		Saturn
9th	Sgr	19h 49m	-21° 53'	-11.4	30'	81%	130° E	+G		
10th	Cap	20h 40m	-20° 14'	-11.7	30'	88%	142° E	+G		

Mercury and Venus

Mercury
5th

Venus
5th

Mercury

Date	Con.	R.A.	Dec.	Mag.	Diam.	Ill.	Elon.	Vis.	Rat.	Close To
1st	Leo	10h 34m	10° 51'	-1.7	5"	99%	2° W	NV	N/A	Venus, Mars, Regulus
3rd	Leo	10h 49m	9° 21'	-1.8	5"	100%	0° E	NV	N/A	Venus, Mars
5th	Leo	11h 3m	7° 49'	-1.7	5"	100%	2° E	NV	N/A	Venus, Mars
7th	Leo	11h 17m	6° 15'	-1.5	5"	99%	4° E	NV	N/A	Venus, Mars
9th	Leo	11h 30m	4° 40'	-1.3	5"	99%	5° E	NV	N/A	Venus, Mars

Venus

Date	Con.	R.A.	Dec.	Mag.	Diam.	Ill.	Elon.	Vis.	Rat.	Close To
1st	Leo	11h 2m	7° 41'	-3.9	10"	100%	5° E	NV	N/A	Mercury, Mars
3rd	Leo	11h 11m	6° 43'	-3.9	10"	100%	6° E	NV	N/A	Mercury, Mars
5th	Leo	11h 20m	5° 45'	-3.9	10"	99%	6° E	NV	N/A	Mercury, Mars
7th	Leo	11h 29m	4° 45'	-3.9	10"	99%	7° E	NV	N/A	Mercury, Mars
9th	Vir	11h 39m	3° 45'	-3.9	10"	99%	7° E	NV	N/A	Mercury, Mars

Mars and the Outer Planets

Mars
5th

Jupiter
5th

Saturn
5th

Mars

Date	Con.	R.A.	Dec.	Mag.	Diam.	Ill.	Elon.	Vis.	Rat.	Close To
1st	Leo	10h 44m	9° 10'	1.7	4"	100%	1° E	NV	N/A	Mercury, Venus, Regulus
5th	Leo	10h 53m	8° 12'	1.7	4"	100%	1° W	NV	N/A	Mercury, Venus
10th	Leo	11h 5m	6° 57'	1.7	4"	100%	2° W	NV	N/A	Mercury, Venus

The Outer Planets

Planet	Date	Con.	R.A.	Dec.	Mag.	Diam.	Elon.	Vis.	Rat.	Close To
Jupiter	5th	Oph	16h 56m	-22° 19'	-2.2	38"	90° E	PM	***	Moon, Antares
Saturn	5th	Sgr	19h 0m	-22° 28'	0.4	18"	121° E	PM	****	
Uranus	5th	Ari	2h 16m	13° 10'	5.7	4"	130° W	AM	***	
Neptune	5th	Aqr	23h 15m	-5° 56'	7.8	2"	175° W	AN	*****	

Highlights

Date	Time (UT)	Event
1st	N/A	The Alpha Aurigid meteor shower is at its maximum. (ZHR: 6)
2nd	07:37	The waxing crescent Moon is north of the bright star Spica. (Virgo, evening sky.)
	N/A	Good opportunity to see Earthshine on the waxing crescent Moon. (Evening sky.)
3rd	19:06	Mars is in conjunction with the Sun. (Not visible.)
4th	01:27	Mercury is at superior conjunction with the Sun. (Not visible.)
5th	19:01	The nearly first quarter Moon is north of the bright star Antares. (Scorpius, evening sky.)
6th	03:11	First Quarter Moon. (Evening sky.)
	06:37	The first quarter Moon is north of Jupiter. (Evening sky.)
8th	12:30	The waxing gibbous Moon is south of Saturn. (Evening sky.)
9th	03:45	The waxing gibbous Moon is south of dwarf planet Pluto. (Evening sky.)
10th	15:57	Neptune is at opposition. (Visible all night.)
	N/A	The Epsilon Perseid meteor shower is at its maximum. (ZHR: 5)

Planet Locations – September 5th

Sun

Mercury

Venus

Mars

Jupiter

Saturn

September 11th – 20th

The Moon

| 11th | 13th | 15th | 17th | 19th |

Date	Con	R.A.	Dec	Mag	Diam	Ill.	Elon.	Phase		Close To
11th	Cap	21h 30m	-17° 43'	-12.0	30'	94%	153° E	+G		
12th	Aqr	22h 17m	-14° 28'	-12.2	29'	97%	164° E	FM		
13th	Aqr	23h 3m	-10° 38'	-12.5	29'	99%	175° E	FM		Neptune
14th	Aqr	23h 48m	-6° 24'	-12.6	29'	100%	175° W	FM		Neptune
15th	Cet	0h 32m	-1° 55'	-12.3	29'	98%	165° W	FM		
16th	Cet	1h 16m	2° 40'	-12.1	30'	95%	155° W	FM		
17th	Psc	2h 1m	7° 12'	-11.8	30'	90%	144° W	-G		Uranus
18th	Ari	2h 47m	11° 30'	-11.5	30'	84%	134° W	-G		Uranus
19th	Tau	3h 35m	15° 23'	-11.1	30'	76%	123° W	-G		Pleiades
20th	Tau	4h 26m	18° 39'	-10.8	31'	67%	111° W	-G		Pleiades, Hyades, Aldebaran

Mercury and Venus

Mercury
15th

Venus
15th

Mercury

Date	Con.	R.A.	Dec.	Mag.	Diam.	Ill.	Elon.	Vis.	Rat.	Close To
11th	Vir	11h 43m	3° 5'	-1.1	5"	98%	7° E	NV	N/A	Venus, Mars
13th	Vir	11h 56m	1° 31'	-0.9	5"	97%	8° E	NV	N/A	Venus
15th	Vir	12h 8m	0° 3'	-0.8	5"	96%	9° E	NV	N/A	Venus
17th	Vir	12h 21m	-1° 36'	-0.7	5"	95%	10° E	NV	N/A	Venus
19th	Vir	12h 32m	-3° 7'	-0.6	5"	93%	12° E	NV	N/A	Venus

Venus

Date	Con.	R.A.	Dec.	Mag.	Diam.	Ill.	Elon.	Vis.	Rat.	Close To
11th	Vir	11h 48m	2° 45'	-3.9	10"	99%	8° E	NV	N/A	Mercury
13th	Vir	11h 57m	1° 44'	-3.9	10"	99%	8° E	NV	N/A	Mercury
15th	Vir	12h 6m	0° 43'	-3.9	10"	99%	9° E	NV	N/A	Mercury
17th	Vir	12h 15m	0° 18'	-3.9	10"	99%	9° E	NV	N/A	Mercury
19th	Vir	12h 24m	-1° 19'	-3.9	10"	99%	10° E	NV	N/A	Mercury

Mars and the Outer Planets

Mars
15th

Jupiter
15th

Saturn
15th

Mars

Date	Con.	R.A.	Dec.	Mag.	Diam.	Ill.	Elon.	Vis.	Rat.	Close To
11th	Leo	11h 8m	6° 42'	1.7	4"	100%	2° W	NV	N/A	Mercury
15th	Leo	11h 17m	5° 42'	1.8	4"	100%	4° W	NV	N/A	
20th	Leo	11h 29m	4° 25'	1.8	4"	100%	5° W	NV	N/A	

The Outer Planets

Planet	Date	Con.	R.A.	Dec.	Mag.	Diam.	Elon.	Vis.	Rat.	Close To
Jupiter	15th	Oph	17h 0m	-22° 26'	-2.1	37"	82° E	PM	**	Antares
Saturn	15th	Sgr	19h 0m	-22° 30'	0.4	17"	112° E	PM	***	
Uranus	15th	Ari	2h 16m	13° 5'	5.7	4"	139° W	AM	***	
Neptune	15th	Aqr	23h 14m	-6° 3'	7.8	2"	176° E	AN	*****	

Highlights

Date	Time (UT)	Event
13th	16:35	The nearly full Moon is south of Neptune. (Visible all night.)
14th	04:34	Full Moon. (Visible all night.)
17th	18:26	The waning gibbous Moon is south of Uranus. (Morning sky.)
18th	04:33	Saturn is stationary prior to resuming prograde motion. (Evening sky.)
19th	17:02	The waning gibbous Moon is south of the Pleiades star cluster. (Taurus, morning sky.)
20th	16:21	The waning gibbous Moon is north of the bright star Aldebaran. (Taurus, morning sky.)

Planet Locations – September 15th

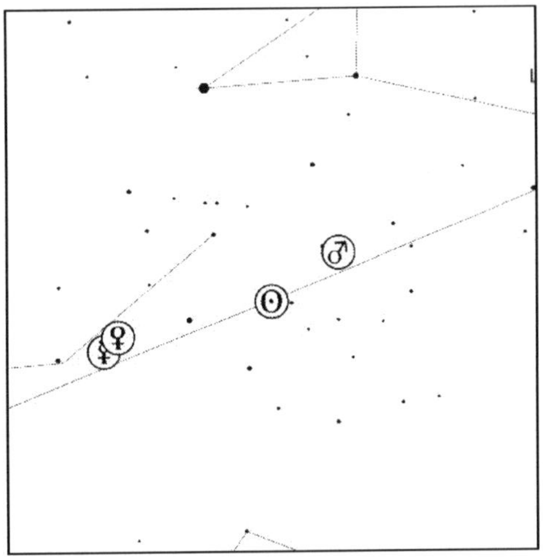

Sun 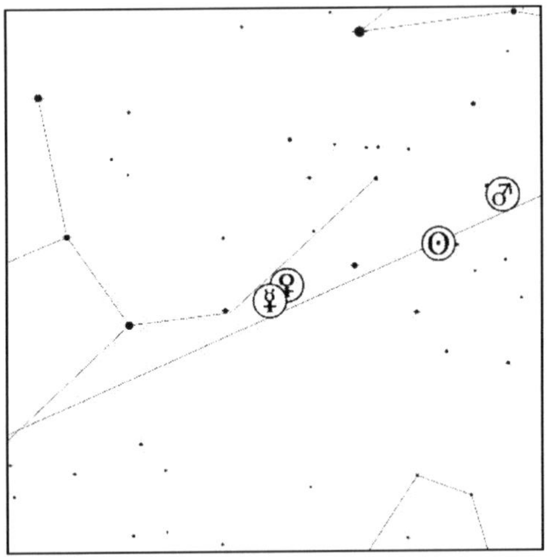 — Mercury 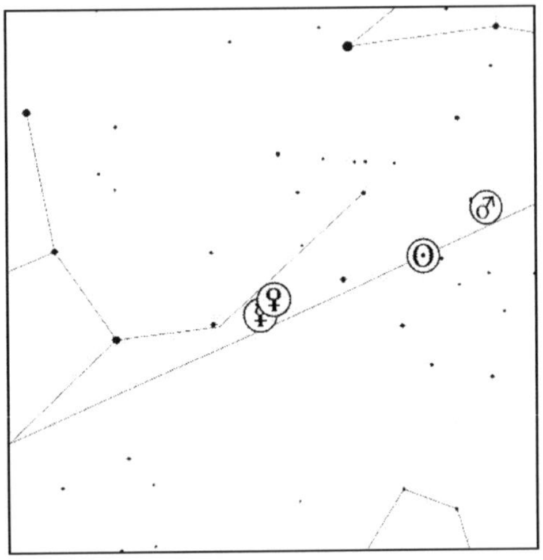 — Venus 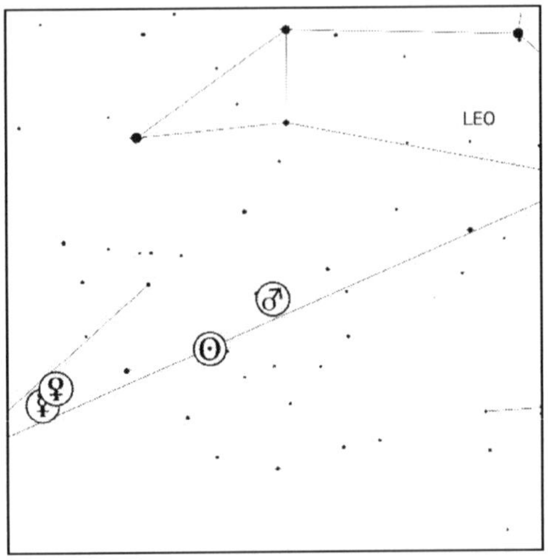 — Mars 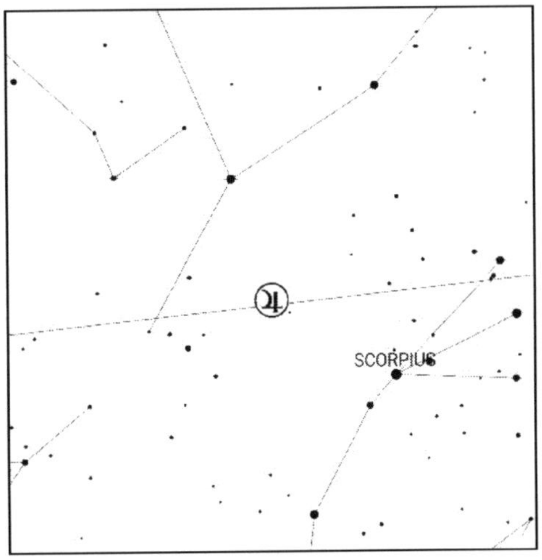 — Jupiter 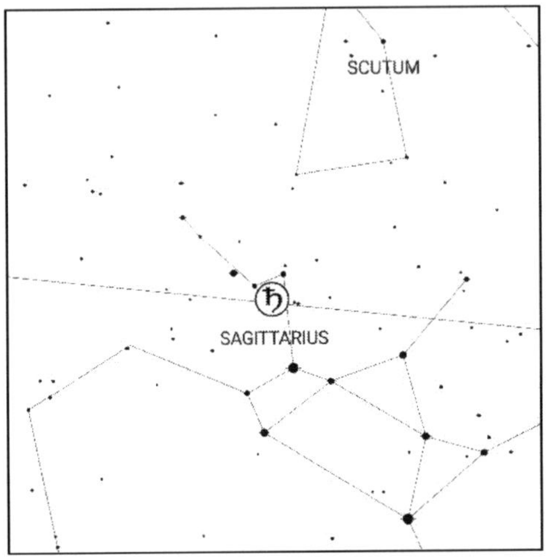 — Saturn

Sun

Mercury

Venus

Mars

LEO

Jupiter

SCORPIUS

Saturn

SCUTUM

SAGITTARIUS

The Moon

| 21st | 23rd | 25th | 27th | 29th |

Date	Con	R.A.	Dec	Mag	Diam	Ill.	Elon.	Phase	Close To
21st	Tau	5h 20m	21° 5'	-10.4	31'	56%	98° W	LQ	Aldebaran
22nd	Gem	6h 17m	22° 28'	-9.9	31'	46%	85° W	LQ	
23rd	Gem	7h 16m	22° 34'	-9.4	32'	35%	71° W	-Cr	
24th	Cnc	8h 16m	21° 16'	-8.8	32'	25%	57° W	-Cr	Praesepe
25th	Cnc	9h 17m	18° 33'	-8.0	33'	15%	43° W	-Cr	Praesepe
26th	Leo	10h 17m	14° 35'	-7.0	33'	8%	29° W	NM	Regulus
27th	Leo	11h 15m	9° 36'	-5.8	33'	3%	15° W	NM	Mars
28th	Vir	12h 12m	3° 59'	-4.6	33'	0%	2° W	NM	Mars
29th	Vir	13h 8m	-1° 51'	-5.1	33'	1%	12° E	NM	Mercury, Venus, Spica
30th	Vir	14h 3m	-7° 29'	-6.4	33'	5%	24° E	NM	Mercury, Spica

Mercury and Venus

Mercury
25th

Venus
25th

Mercury

Date	Con.	R.A.	Dec.	Mag.	Diam.	Ill.	Elon.	Vis.	Rat.	Close To
21st	Vir	12h 44m	-4° 36'	-0.5	5"	92%	13° E	NV	N/A	Venus
23rd	Vir	12h 55m	-6° 4'	-0.4	5"	91%	14° E	NV	N/A	Venus, Spica
25th	Vir	13h 7m	-7° 29'	-0.3	5"	90%	15° E	NV	N/A	Venus, Spica
27th	Vir	13h 18m	-8° 52'	-0.3	5"	88%	16° E	PM	**	Venus, Spica
29th	Vir	13h 29m	-10° 13'	-0.2	5"	87%	17° E	PM	**	Moon, Venus, Spica

Venus

Date	Con.	R.A.	Dec.	Mag.	Diam.	Ill.	Elon.	Vis.	Rat.	Close To
21st	Vir	12h 33m	-2° 20'	-3.9	10"	98%	10° E	NV	N/A	Mercury
23rd	Vir	12h 42m	-3° 21'	-3.9	10"	98%	10° E	NV	N/A	Mercury
25th	Vir	12h 51m	-4° 22'	-3.9	10"	98%	11° E	NV	N/A	Mercury, Spica
27th	Vir	13h 0m	-5° 23'	-3.9	10"	98%	11° E	NV	N/A	Mercury, Spica
29th	Vir	13h 9m	-6° 23'	-3.9	10"	98%	12° E	NV	N/A	Moon, Mercury, Spica

Mars and the Outer Planets

Mars
25th

Jupiter
25th

Saturn
25th

Mars

Date	Con.	R.A.	Dec.	Mag.	Diam.	Ill.	Elon.	Vis.	Rat.	Close To
21st	Leo	11h 31m	4° 10'	1.8	4"	100%	5° W	NV	N/A	
25th	Vir	11h 41m	3° 8'	1.8	4"	100%	7° W	NV	N/A	
30th	Vir	11h 52m	1° 51'	1.8	4"	100%	8° W	NV	N/A	

The Outer Planets

Planet	Date	Con.	R.A.	Dec.	Mag.	Diam.	Elon.	Vis.	Rat.	Close To
Jupiter	25th	Oph	17h 5m	-22° 34'	-2.1	36"	74° E	PM	**	Antares
Saturn	25th	Sgr	19h 0m	-22° 31'	0.4	17"	103° E	PM	***	
Uranus	25th	Ari	2h 14m	12° 59'	5.7	4"	148° W	AM	***	
Neptune	25th	Aqr	23h 13m	-6° 9'	7.8	2"	166° E	PM	*****	

Highlights

Date	Time (UT)	Event
22nd	02:42	Last Quarter Moon. (Morning sky.)
23rd	07:51	Autumn Equinox.
24th	21:00	The waning crescent Moon is south of the Praesepe star cluster. (Cancer, morning sky.)
	23:32	Asteroid Vesta is stationary prior to beginning retrograde motion. (Morning sky.)
25th	N/A	Good opportunity to see Earthshine on the waning crescent Moon. (Morning sky.)
26th	07:52	The waning crescent Moon is north of the bright star Regulus. (Leo, morning sky.)
28th	17:58	Mercury is 1.4° north of the bright star Spica. (Virgo, evening sky.)
	18:27	New Moon. (Not visible.)
29th	22:50	The just-past new Moon is north of Mercury. (Evening sky.)

Planet Locations – September 25th

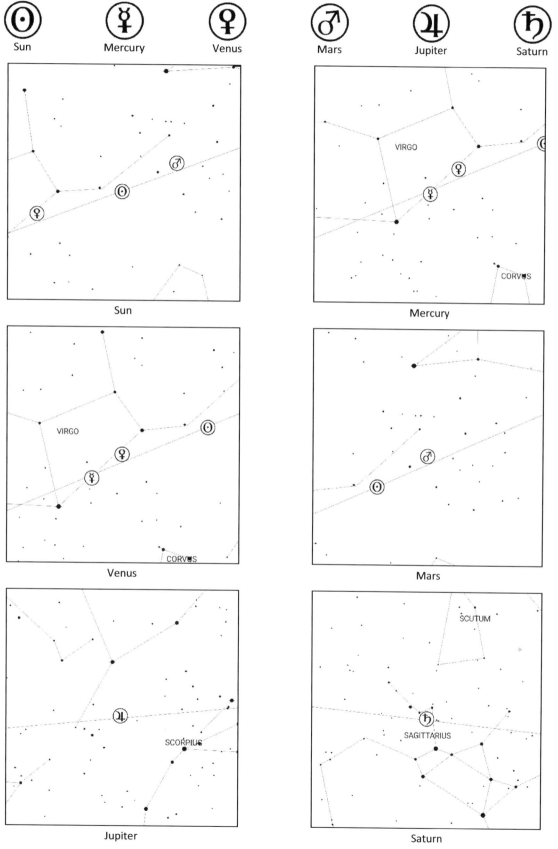

October 1st to 10th

The Moon

| | 1st | 3rd | 5th | 7th | 9th |

Date	Con	R.A.	Dec	Mag	Diam	Ill.	Elon.	Phase	Close To
1st	Lib	14h 59m	-12° 35'	-7.4	32'	11%	37° E	NM	
2nd	Lib	15h 54m	-16° 49'	-8.3	32'	19%	50° E	+Cr	Antares
3rd	Oph	16h 50m	-19° 59'	-9.0	31'	28%	64° E	+Cr	Jupiter, Antares
4th	Sgr	17h 46m	-21° 59'	-9.6	31'	38%	77° E	FQ	Jupiter
5th	Sgr	18h 41m	-22° 46'	-10.0	30'	48%	89° E	FQ	Saturn
6th	Sgr	19h 35m	-22° 24'	-10.5	30'	58%	102° E	FQ	Saturn
7th	Cap	20h 27m	-20° 58'	-10.8	30'	67%	114° E	+G	
8th	Cap	21h 17m	-18° 37'	-11.2	30'	76%	126° E	+G	
9th	Aqr	22h 5m	-15° 30'	-11.5	29'	83%	137° E	+G	
10th	Aqr	22h 51m	-11° 46'	-11.8	29'	90%	147° E	+G	Neptune

Mercury and Venus

Mercury
5th

Venus
5th

Mercury

Date	Con.	R.A.	Dec.	Mag.	Diam.	Ill.	Elon.	Vis.	Rat.	Close To
1st	Vir	13h 39m	-11° 31'	-0.2	5"	85%	18° E	PM	**	Venus, Spica
3rd	Vir	13h 50m	-12° 46'	-0.2	5"	83%	18° E	PM	**	Venus, Spica
5th	Vir	14h 1m	-13° 58'	-0.1	5"	82%	19° E	PM	**	Venus, Spica
7th	Vir	14h 11m	-15° 6'	-0.1	6"	80%	20° E	PM	***	Venus
9th	Vir	14h 21m	-16° 12'	-0.1	6"	78%	21° E	PM	***	Venus

Venus

Date	Con.	R.A.	Dec.	Mag.	Diam.	Ill.	Elon.	Vis.	Rat.	Close To
1st	Vir	13h 19m	-7° 22'	-3.9	10"	97%	12° E	NV	N/A	Mercury, Spica
3rd	Vir	13h 28m	-8° 21'	-3.9	10"	97%	13° E	NV	N/A	Mercury, Spica
5th	Vir	13h 37m	-9° 19'	-3.9	10"	97%	13° E	NV	N/A	Mercury, Spica
7th	Vir	13h 46m	-10° 17'	-3.9	10"	97%	14° E	NV	N/A	Mercury, Spica
9th	Vir	13h 56m	-11° 13'	-3.9	10"	97%	14° E	NV	N/A	Mercury, Spica

Mars and the Outer Planets

Mars
5th

Jupiter
5th

Saturn
5th

Mars

Date	Con.	R.A.	Dec.	Mag.	Diam.	Ill.	Elon.	Vis.	Rat.	Close To
1st	Vir	11h 55m	1° 35'	1.8	4"	100%	9° W	NV	N/A	
5th	Vir	12h 4m	0° 33'	1.8	4"	100%	10° W	NV	N/A	
10th	Vir	12h 16m	0° 45'	1.8	4"	100%	11° W	NV	N/A	

The Outer Planets

Planet	Date	Con.	R.A.	Dec.	Mag.	Diam.	Elon.	Vis.	Rat.	Close To
Jupiter	5th	Oph	17h 11m	-22° 43'	-2.0	35"	67° E	PM	**	
Saturn	5th	Sgr	19h 1m	-22° 31'	0.5	17"	94° E	PM	***	Moon
Uranus	5th	Ari	2h 13m	12° 52'	5.7	4"	158° W	AM	***	
Neptune	5th	Aqr	23h 12m	-6° 15'	7.8	2"	157° E	PM	*****	

Highlights

Date	Time (UT)	Event
2nd	12:44	Dwarf planet Pluto is stationary prior to resuming prograde motion. (Evening sky.)
	N/A	Good opportunity to see Earthshine on the waxing crescent Moon. (Evening sky.)
3rd	03:01	The waxing crescent Moon is north of the bright star Antares. (Scorpius, evening sky.)
	21:37	The waxing crescent Moon is north of Jupiter. (Evening sky.)
5th	16:48	First Quarter Moon. (Evening sky.)
	21:38	The first quarter Moon is south of Saturn. (Evening sky.)
6th	08:09	The just-past first quarter Moon is south of dwarf planet Pluto. (Evening sky.)
8th	N/A	The Darconid meteor shower is at its maximum. (ZHR: Variable.)
10th	23:10	The waxing gibbous Moon is south of Neptune. (Evening sky.)
	N/A	The Southern Taurid meteor shower is at its maximum. (ZHR: 5)

Planet Locations – October 5th

| ☉ Sun | ☿ Mercury | ♀ Venus | ♂ Mars | ♃ Jupiter | ♄ Saturn |

Sun

Mercury

Venus

Mars

Jupiter

Saturn

The Moon

| 11th | 13th | 15th | 17th | 19th |

Date	Con	R.A.	Dec	Mag	Diam	Ill.	Elon.	Phase		Close To
11th	Aqr	23h 36m	-7° 35'	-12.0	29'	95%	158° E	+G		Neptune
12th	Psc	0h 21m	-3° 6'	-12.3	30'	98%	168° E	FM		
13th	Cet	1h 5m	1° 33'	-12.6	30'	100%	178° E	FM		
14th	Psc	1h 50m	6° 11'	-12.5	30'	99%	172° W	FM		Uranus
15th	Ari	2h 36m	10° 38'	-12.2	30'	97%	161° W	FM		Uranus
16th	Ari	3h 24m	14° 42'	-12.0	30'	93%	150° W	-G		Pleiades
17th	Tau	4h 14m	18° 11'	-11.6	30'	88%	138° W	-G		Pleiades, Hyades, Aldebaran
18th	Tau	5h 7m	20° 51'	-11.3	31'	80%	126° W	-G		Hyades, Aldebaran
19th	Gem	6h 3m	22° 30'	-11.0	31'	71%	113° W	-G		
20th	Gem	7h 0m	22° 56'	-10.6	31'	61%	100° W	LQ		

Mercury and Venus

Mercury
15th

Venus
15th

Mercury

Date	Con.	R.A.	Dec.	Mag.	Diam.	Ill.	Elon.	Vis.	Rat.		Close To
11th	Lib	14h 31m	-17° 13'	-0.1	6"	75%	21° E	PM	***		Venus
13th	Lib	14h 41m	-18° 10'	-0.1	6"	73%	22° E	PM	***		Venus
15th	Lib	14h 50m	-19° 3'	-0.1	6"	70%	22° E	PM	***		Venus
17th	Lib	14h 59m	-19° 51'	-0.1	6"	67%	23° E	PM	***		Venus
19th	Lib	15h 8m	-20° 34'	-0.1	7"	63%	23° E	PM	***		Venus

Venus

Date	Con.	R.A.	Dec.	Mag.	Diam.	Ill.	Elon.	Vis.	Rat.	Close To
11th	Vir	14h 5m	-12° 9'	-3.9	10"	96%	15° E	NV	N/A	Mercury
13th	Vir	14h 15m	-13° 3'	-3.9	10"	96%	15° E	PM	*	Mercury
15th	Lib	14h 24m	-13° 56'	-3.9	10"	96%	16° E	PM	*	Mercury
17th	Lib	14h 34m	-14° 48'	-3.9	10"	96%	17° E	PM	*	Mercury
19th	Lib	14h 44m	-15° 38'	-3.9	10"	95%	17° E	PM	*	Mercury

Mars and the Outer Planets

Mars
15th

Jupiter
15th

Saturn
15th

Mars

Date	Con.	R.A.	Dec.	Mag.	Diam.	Ill.	Elon.	Vis.	Rat.	Close To
11th	Vir	12h 18m	-1° 1'	1.8	4"	100%	12° W	NV	N/A	
15th	Vir	12h 28m	-2° 3'	1.8	4"	99%	13° W	NV	N/A	
20th	Vir	12h 40m	-3° 21'	1.8	4"	99%	15° W	NV	N/A	

The Outer Planets

Planet	Date	Con.	R.A.	Dec.	Mag.	Diam.	Elon.	Vis.	Rat.	Close To
Jupiter	15th	Oph	17h 18m	-22° 52'	-2.0	35"	59° E	PM	**	
Saturn	15th	Sgr	19h 2m	-22° 29'	0.5	16"	86° E	PM	***	
Uranus	15th	Ari	2h 12m	12° 45'	5.7	4"	167° W	AM	****	Moon
Neptune	15th	Aqr	23h 11m	-6° 20'	7.8	2"	148° E	PM	****	

Highlights

Date	Time (UT)	Event
12th	N/A	The Delta Aquariid meteor shower is at its maximum. (ZHR: 2)
13th	21:09	Full Moon. (Visible all night.)
14th	23:05	The just-past full Moon is south of Uranus. (Morning sky.)
16th	21:51	The waning gibbous Moon is south of the Pleiades star cluster. (Taurus, morning sky.)
17th	20:32	The waning gibbous Moon is north of the bright star Aldebaran. (Taurus, morning sky.)
18th	N/A	The Epsilon Geminid meteor shower is at its maximum. (ZHR: 3)
20th	03:47	Mercury is at greatest eastern elongation from the Sun. (Evening sky.)

Planet Locations – October 15th

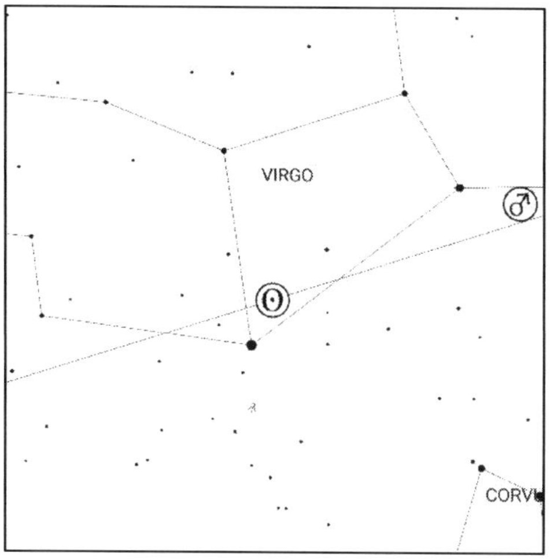

☉ Sun	☿ Mercury	♀ Venus	♂ Mars	♃ Jupiter	♄ Saturn

VIRGO

♂

☉

CORVUS

Sun

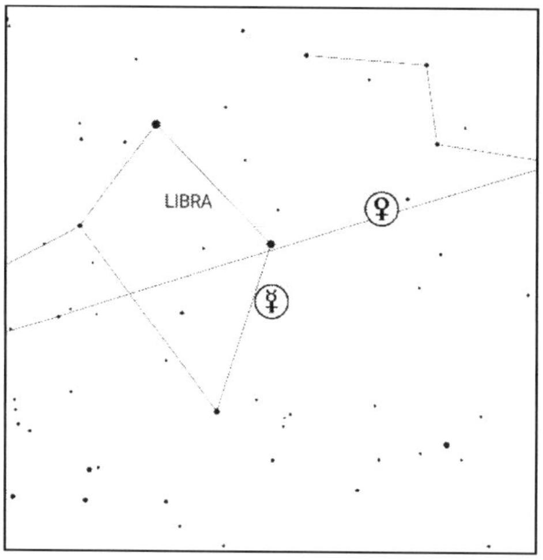

LIBRA

♀

☿

Mercury

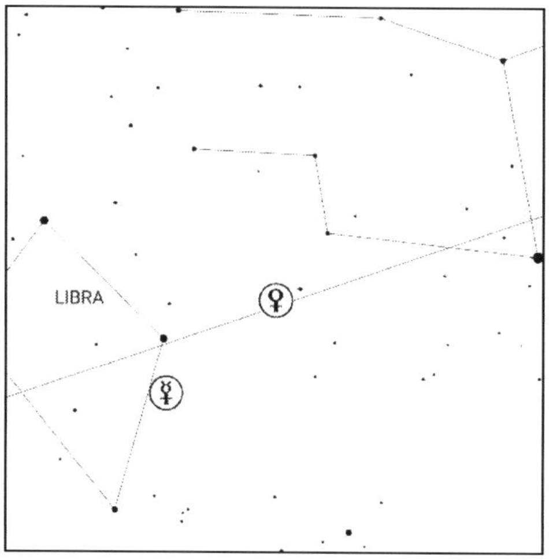

LIBRA

♀

☿

Venus

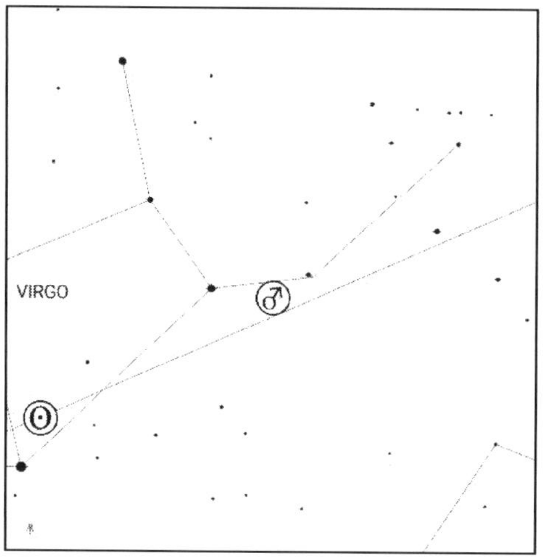

VIRGO

♂

☉

Mars

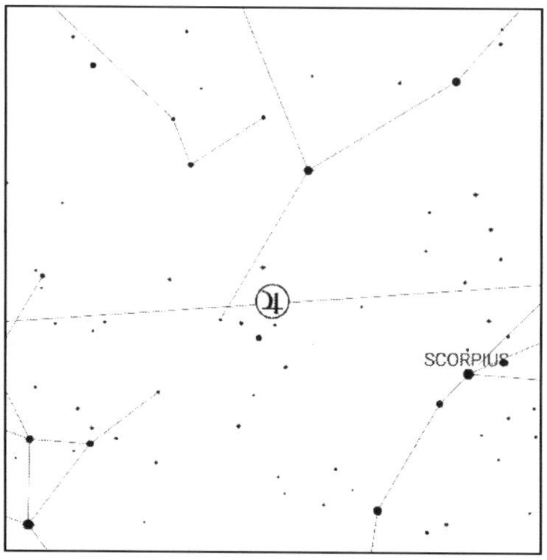

♃

SCORPIUS

Jupiter

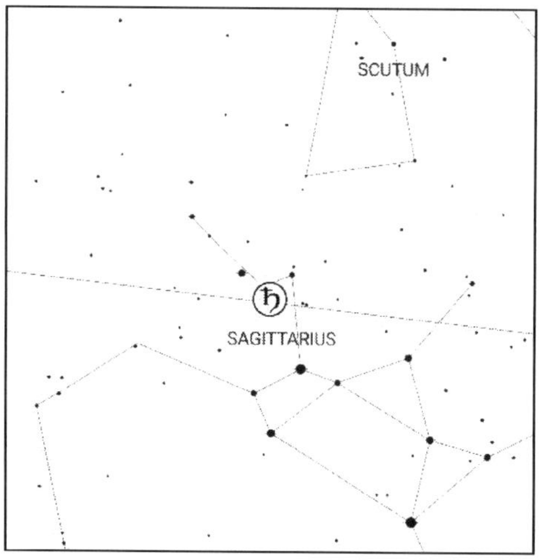

SCUTUM

♄

SAGITTARIUS

Saturn

October 21st – 31st

The Moon

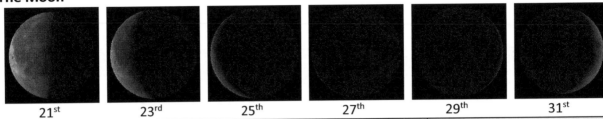

| 21st | 23rd | 25th | 27th | 29th | 31st |

Date	Con	R.A.	Dec	Mag	Diam	Ill.	Elon.	Phase	Close To
21st	Gem	7h 59m	22° 2'	-10.1	32'	50%	86° W	LQ	Praesepe
22nd	Cnc	8h 57m	19° 49'	-9.6	32'	39%	72° W	LQ	Praesepe
23rd	Leo	9h 55m	16° 20'	-9.0	33'	28%	59° W	-Cr	Regulus
24th	Leo	10h 52m	11° 48'	-8.3	33'	18%	45° W	-Cr	Regulus
25th	Vir	11h 48m	6° 30'	-7.4	33'	10%	32° W	NM	
26th	Vir	12h 43m	0° 46'	-6.3	33'	4%	20° W	NM	Mars, Spica
27th	Vir	13h 38m	-5° 1'	-5.0	33'	1%	7° W	NM	Mars, Spica
28th	Lib	14h 34m	-10° 28'	-4.6	33'	0%	6° E	NM	
29th	Lib	15h 30m	-15° 14'	-5.8	32'	3%	19° E	NM	Mercury, Venus
30th	Oph	16h 27m	-19° 0'	-6.9	32'	7%	32° E	NM	Antares
31st	Oph	17h 25m	-21° 35'	-7.8	31'	14%	46° E	+Cr	Jupiter

Mercury and Venus

Mercury
25th

Venus
25th

Mercury

Date	Con.	R.A.	Dec.	Mag.	Diam.	Ill.	Elon.	Vis.	Rat.	Close To
21st	Lib	15h 15m	-21° 11'	0.0	7"	59%	23° E	PM	***	Venus
23rd	Lib	15h 23m	-21° 42'	0.0	7"	55%	23° E	PM	***	Venus
25th	Lib	15h 29m	-22° 5'	0.1	7"	50%	23° E	PM	***	Venus
27th	Lib	15h 33m	-22° 20'	0.2	8"	44%	22° E	PM	***	Venus
29th	Lib	15h 37m	-22° 25'	0.3	8"	38%	21° E	PM	***	Moon, Venus
31st	Lib	15h 38m	-22° 19'	0.6	8"	31%	19° E	PM	***	Venus

Venus

Date	Con.	R.A.	Dec.	Mag.	Diam.	Ill.	Elon.	Vis.	Rat.	Close To
21st	Lib	14h 53m	-16° 27'	-3.9	10"	95%	18° E	PM	*	Mercury
23rd	Lib	15h 3m	-17° 14'	-3.9	10"	95%	18° E	PM	*	Mercury
25th	Lib	15h 13m	-17° 59'	-3.9	11"	95%	19° E	PM	*	Mercury
27th	Lib	15h 23m	-18° 43'	-3.9	11"	94%	19° E	PM	*	Mercury
29th	Lib	15h 33m	-19° 24'	-3.9	11"	94%	20° E	PM	*	Moon, Mercury
31st	Lib	15h 44m	-20° 3'	-3.9	11"	94%	21° E	PM	*	Mercury

Mars and the Outer Planets

Mars	Jupiter	Saturn
25th	25th	25th

Mars

Date	Con.	R.A.	Dec.	Mag.	Diam.	Ill.	Elon.	Vis.	Rat.	Close To
21st	Vir	12h 42m	-3° 37'	1.8	4"	99%	15° W	AM	*	
25th	Vir	12h 52m	-4° 38'	1.8	4"	99%	17° W	AM	*	Spica
31st	Vir	13h 6m	-6° 10'	1.8	4"	99%	19° W	AM	*	Spica

The Outer Planets

Planet	Date	Con.	R.A.	Dec.	Mag.	Diam.	Elon.	Vis.	Rat.	Close To
Jupiter	25th	Oph	17h 25m	-22° 60'	-1.9	34"	52° E	PM	**	
Saturn	25th	Sgr	19h 5m	-22° 27'	0.6	16"	77° E	PM	***	
Uranus	25th	Ari	2h 10m	12° 36'	5.7	4"	177° W	AN	****	
Neptune	25th	Aqr	23h 10m	-6° 24'	7.8	2"	138° E	PM	****	

Highlights

Date	Time (UT)	Event
21st	12:40	Last Quarter Moon. (Morning sky.)
	N/A	The Orionid meteor shower is at its maximum. (ZHR: 25)
22nd	03:50	The just-past last quarter Moon is north of the Praesepe star cluster. (Cancer, morning sky.)
23rd	17:50	The waning crescent Moon is north of the bright star Regulus. (Leo, morning sky.)
25th	N/A	Good opportunity to see Earthshine on the waning crescent Moon. (Morning sky.)
26th	18:06	The waning crescent Moon is north of Mars. (Morning sky.)
28th	03:39	New Moon. (Not visible.)
	12:07	Uranus is at opposition. (Visible all night.)
30th	08:15	Mercury is 2.7° south of Venus. (Evening sky.)
31st	14:11	The waxing crescent Moon is north of Jupiter. (Evening sky.)
31st	20:25	Mercury is stationary prior to beginning retrograde motion. (Evening sky.)
	N/A	Good opportunity to see Earthshine on the waxing crescent Moon. (Evening sky.)

Planet Locations – October 25th

Sun · Mercury · Venus · Mars · Jupiter · Saturn

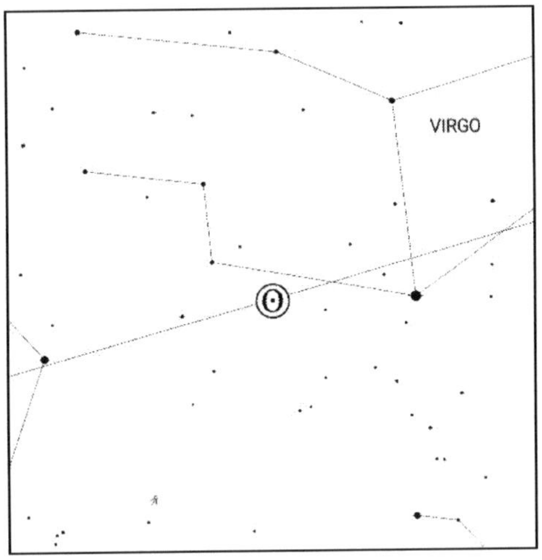

Sun

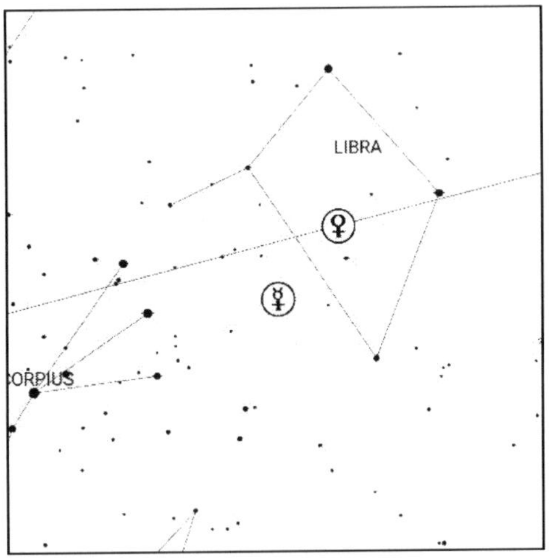

Mercury

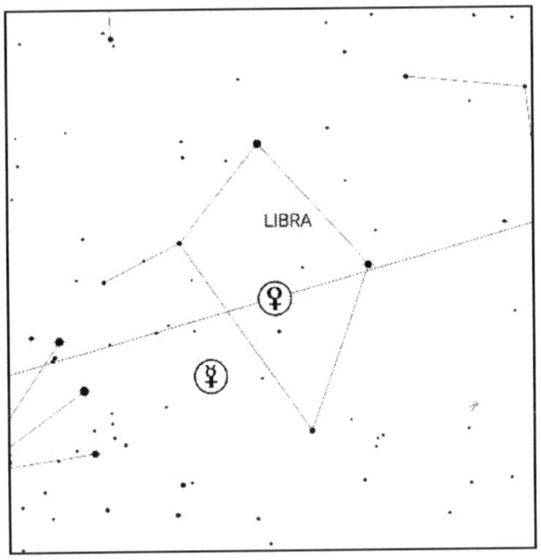

Venus

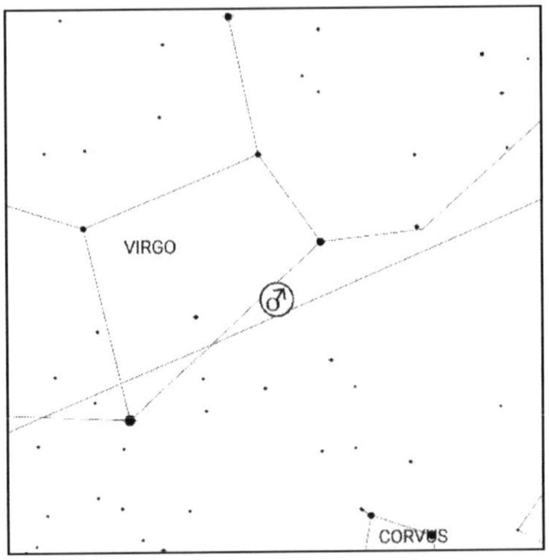

Mars

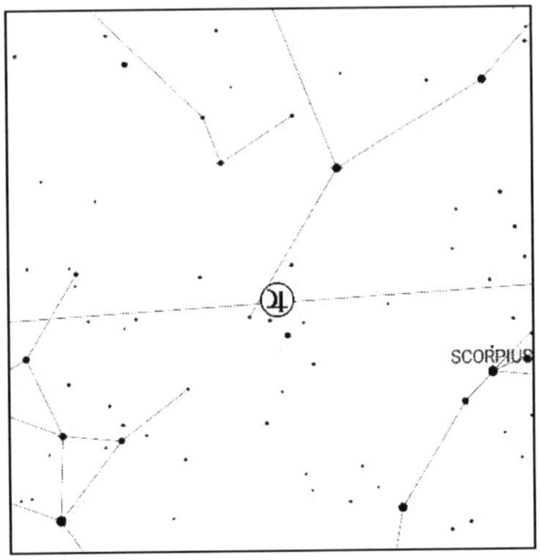

Jupiter

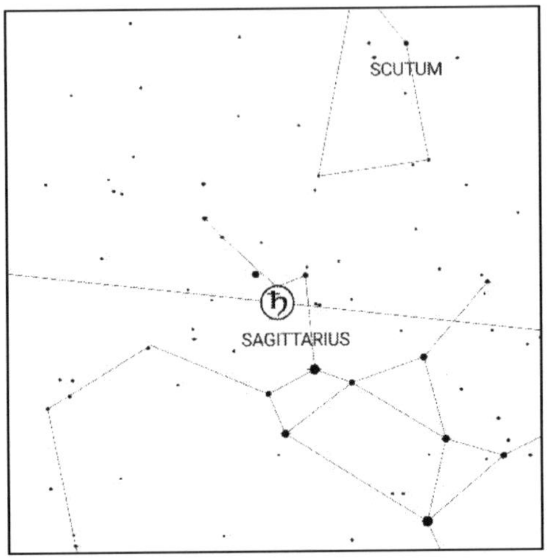

Saturn

The Moon

| 1st | 3rd | 5th | 7th | 9th |

Date	Con	R.A.	Dec	Mag	Diam	Ill.	Elon.	Phase	Close To
1st	Sgr	18h 22m	-22° 53'	-8.6	31'	22%	59° E	+Cr	Saturn
2nd	Sgr	19h 17m	-22° 54'	-9.2	31'	31%	72° E	+Cr	Saturn
3rd	Cap	20h 11m	-21° 47'	-9.7	30'	41%	85° E	FQ	
4th	Cap	21h 3m	-19° 40'	-10.2	30'	51%	96° E	FQ	
5th	Cap	21h 52m	-16° 43'	-10.5	30'	60%	108° E	FQ	
6th	Aqr	22h 38m	-13° 7'	-10.9	30'	69%	118° E	+G	Neptune
7th	Aqr	23h 24m	-9° 1'	-11.2	29'	77%	129° E	+G	Neptune
8th	Psc	0h 8m	-4° 34'	-11.5	30'	85%	139° E	+G	
9th	Cet	0h 52m	0° 6'	-11.8	30'	91%	149° E	+G	
10th	Psc	1h 37m	4° 48'	-12.1	30'	96%	159° E	FM	Uranus

Mercury and Venus

Mercury
5th

Venus
5th

Mercury

Date	Con.	R.A.	Dec.	Mag.	Diam.	Ill.	Elon.	Vis.	Rat.	Close To
1st	Lib	15h 38m	-22° 11'	0.7	9"	27%	18° E	PM	***	Venus
3rd	Lib	15h 36m	-21° 45'	1.2	9"	20%	16° E	PM	**	Venus
5th	Lib	15h 32m	-21° 3'	1.8	9"	13%	13° E	NV	N/A	Venus
7th	Lib	15h 25m	-20° 5'	2.7	10"	6%	9° E	NV	N/A	
9th	Lib	15h 16m	-18° 53'	4.0	10"	2%	5° E	NV	N/A	

Venus

Date	Con.	R.A.	Dec.	Mag.	Diam.	Ill.	Elon.	Vis.	Rat.	Close To
1st	Lib	15h 49m	-20° 22'	-3.9	11"	94%	21° E	PM	*	Mercury, Antares
3rd	Sco	15h 59m	-20° 58'	-3.9	11"	93%	22° E	PM	*	Mercury, Antares
5th	Sco	16h 9m	-21° 32'	-3.9	11"	93%	22° E	PM	*	Mercury, Antares
7th	Sco	16h 20m	-22° 3'	-3.9	11"	93%	23° E	PM	*	Antares
9th	Oph	16h 30m	-22° 32'	-3.9	11"	93%	23° E	PM	*	Antares

Mars and the Outer Planets

Mars
5th

Jupiter
5th

Saturn
5th

Mars

Date	Con.	R.A.	Dec.	Mag.	Diam.	Ill.	Elon.	Vis.	Rat.	Close To
1st	Vir	13h 8m	-6° 25'	1.8	4"	99%	19° W	AM	*	Spica
5th	Vir	13h 18m	-7° 26'	1.8	4"	99%	21° W	AM	*	Spica
10th	Vir	13h 30m	-8° 40'	1.8	4"	99%	23° W	AM	*	Spica

The Outer Planets

Planet	Date	Con.	R.A.	Dec.	Mag.	Diam.	Elon.	Vis.	Rat.	Close To
Jupiter	5th	Oph	17h 34m	-23° 8'	-1.9	33"	43° E	PM	*	
Saturn	5th	Sgr	19h 8m	-22° 23'	0.6	16"	67° E	PM	**	
Uranus	5th	Ari	2h 8m	12° 27'	5.7	4"	172° E	AN	****	
Neptune	5th	Aqr	23h 10m	-6° 28'	7.9	2"	127° E	PM	****	

Highlights

Date	Time (UT)	Event
2nd	06:40	The waxing crescent Moon is south of Saturn. (Evening sky.)
	17:15	The waxing crescent Moon is south of dwarf planet Pluto. (Evening sky.)
4th	10:24	First quarter Moon. (Evening sky.)
7th	05:31	The waxing gibbous Moon is south of Neptune. (Evening sky.)
8th	04:16	Mars is 3.1° north of the bright star Spica. (Virgo, morning sky.)
9th	04:57	Venus is 4.0° north of the bright star Antares. (Scorpius, evening sky.)

Planet Locations – November 5th

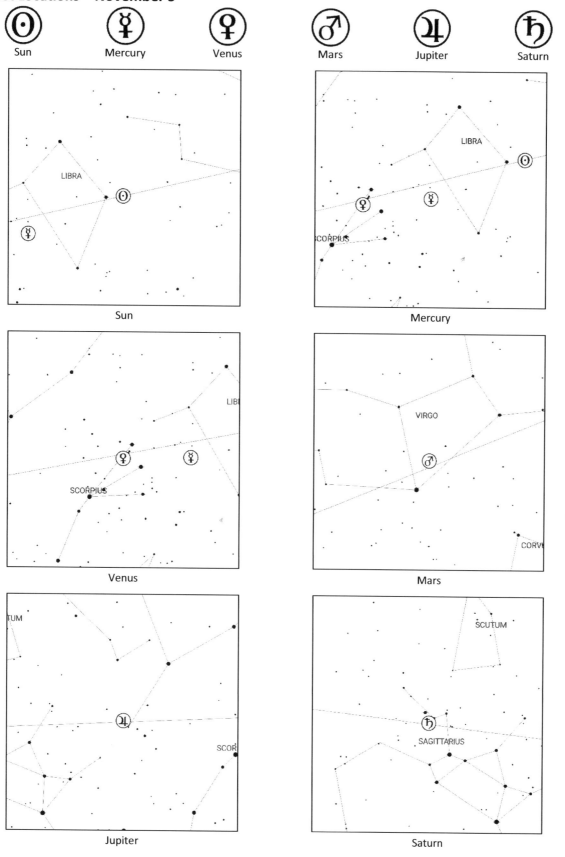

November 11th – 20th

The Moon

| 11th | 13th | 15th | 17th | 19th |

Date	Con	R.A.	Dec	Mag	Diam	Ill.	Elon.	Phase	Close To
11th	Cet	2h 22m	9° 24'	-12.4	30'	99%	169° E	FM	Uranus
12th	Ari	3h 10m	13° 41'	-12.6	30'	100%	180° W	FM	Pleiades
13th	Tau	4h 1m	17° 26'	-12.4	31'	99%	168° W	FM	Pleiades, Hyades, Aldebaran
14th	Tau	4h 54m	20° 25'	-12.1	31'	96%	156° W	FM	Hyades, Aldebaran
15th	Tau	5h 49m	22° 23'	-11.8	31'	91%	143° W	-G	
16th	Gem	6h 47m	23° 9'	-11.5	31'	84%	130° W	-G	
17th	Gem	7h 46m	22° 35'	-11.1	32'	75%	116° W	-G	
18th	Cnc	8h 44m	20° 40'	-10.7	32'	65%	102° W	LQ	Praesepe
19th	Leo	9h 41m	17° 32'	-10.3	32'	54%	89° W	LQ	Regulus
20th	Leo	10h 37m	13° 21'	-9.8	32'	43%	76° W	LQ	Regulus

Mercury and Venus

Mercury
15th

Venus
15th

Mercury

Date	Con.	R.A.	Dec.	Mag.	Diam.	Ill.	Elon.	Vis.	Rat.	Close To
11th	Lib	15h 6m	-17° 32'	5.7	10"	0%	0° E	NV	N/A	
13th	Lib	14h 56m	-16° 9'	4.2	10"	1%	4° W	NV	N/A	
15th	Lib	14h 48m	-14° 55'	2.7	9"	6%	8° W	NV	N/A	
17th	Lib	14h 42m	-13° 57'	1.6	9"	14%	12° W	NV	N/A	
19th	Lib	14h 39m	-13° 20'	0.8	9"	23%	15° W	NV	N/A	

Venus

Date	Con.	R.A.	Dec.	Mag.	Diam.	Ill.	Elon.	Vis.	Rat.	Close To
11th	Oph	16h 41m	-22° 58'	-3.9	11"	92%	24° E	PM	*	Antares
13th	Oph	16h 52m	-23° 22'	-3.9	11"	92%	25° E	PM	*	Antares
15th	Oph	17h 2m	-23° 42'	-3.9	11"	92%	25° E	PM	*	Antares
17th	Oph	17h 13m	-24° 0'	-3.9	11"	91%	26° E	PM	*	Jupiter
19th	Oph	17h 24m	-24° 16'	-3.9	11"	91%	27° E	PM	*	Jupiter

Mars and the Outer Planets

Mars
15th

Jupiter
15th

Saturn
15th

Mars

Date	Con.	R.A.	Dec.	Mag.	Diam.	Ill.	Elon.	Vis.	Rat.	Close To
11th	Vir	13h 33m	-8° 55'	1.8	4"	98%	23° W	AM	*	Spica
15th	Vir	13h 43m	-9° 53'	1.8	4"	98%	25° W	AM	*	Spica
20th	Vir	13h 55m	-11° 5'	1.7	4"	98%	27° W	AM	*	Spica

The Outer Planets

Planet	Date	Con.	R.A.	Dec.	Mag.	Diam.	Elon.	Vis.	Rat.	Close To
Jupiter	15th	Oph	17h 43m	-23° 13'	-1.9	33"	36° E	PM	*	
Saturn	15th	Sgr	19h 11m	-22° 18'	0.6	16"	58° E	PM	**	
Uranus	15th	Ari	2h 7m	12° 20'	5.7	4"	161° E	PM	****	
Neptune	15th	Aqr	23h 9m	-6° 30'	7.9	2"	117° E	PM	****	

Highlights

Date	Time (UT)	Event
11th	05:36	The nearly full Moon is south of Uranus. (Evening sky.)
	12:35	Transit of Mercury across the Sun begins.
	15:15	Mercury is at inferior conjunction with the Sun. (Not visible.)
	15:20	Mercury at mid transit across the Sun.
	18:04	Transit of Mercury across the Sun ends.
12th	13:35	Full Moon. (Visible all night.)
13th	06:41	The just-past full Moon is south of the Pleiades star cluster. (Taurus, visible all night.)
14th	04:55	The waning gibbous Moon is north of the bright star Aldebaran. (Taurus, morning sky.)
18th	11:18	The nearly last quarter Moon is north of the Praesepe star cluster. (Cancer, morning sky.)
	N/A	The Leonid meteor shower is at its maximum. (ZHR: 20)
19th	21:12	Last Quarter Moon. (Morning sky.)
	22:17	The last quarter Moon is north of the bright star Regulus. (Leo, morning sky.)
20th	14:19	Mercury is stationary prior to resuming prograde motion. (Morning sky.)

| Sun | Mercury | Venus | Mars | Jupiter | Saturn |

Sun

Mercury

Venus

Mars

Jupiter

Saturn

The Moon

  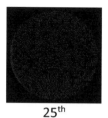  

21st 23rd 25th 27th 29th

Date	Con	R.A.	Dec	Mag	Diam	Ill.	Elon.	Phase	Close To
21st	Leo	11h 31m	8° 22'	-9.2	32'	32%	64° W	-Cr	
22nd	Vir	12h 25m	2° 54'	-8.5	33'	22%	51° W	-Cr	
23rd	Vir	13h 18m	-2° 47'	-7.7	33'	13%	39° W	-Cr	Mars, Spica
24th	Vir	14h 12m	-8° 18'	-6.7	33'	6%	27° W	NM	Mercury, Mars, Spica
25th	Lib	15h 6m	-13° 21'	-5.5	32'	2%	14° W	NM	Mercury
26th	Lib	16h 3m	-17° 35'	-4.3	32'	0%	1° W	NM	Antares
27th	Oph	17h 0m	-20° 44'	-5.1	32'	1%	12° E	NM	Antares
28th	Sgr	17h 58m	-22° 37'	-6.3	31'	4%	26° E	NM	Venus, Jupiter
29th	Sgr	18h 56m	-23° 12'	-7.3	31'	10%	39° E	NM	Venus, Saturn
30th	Sgr	19h 51m	-22° 31'	-8.1	31'	16%	52° E	+Cr	Saturn

Mercury and Venus

Mercury
25th

Venus
25th

Mercury

Date	Con.	R.A.	Dec.	Mag.	Diam.	Ill.	Elon.	Vis.	Rat.	Close To
21st	Lib	14h 39m	-13° 5'	0.3	8"	32%	17° W	AM	***	
23rd	Lib	14h 42m	-13° 9'	-0.1	8"	42%	18° W	AM	***	Mars
25th	Lib	14h 46m	-13° 29'	-0.3	7"	50%	19° W	AM	***	Moon, Mars
27th	Lib	14h 53m	-14° 2'	-0.4	7"	58%	20° W	AM	***	Mars
29th	Lib	15h 1m	-14° 43'	-0.5	7"	65%	20° W	AM	***	

Venus

Date	Con.	R.A.	Dec.	Mag.	Diam.	Ill.	Elon.	Vis.	Rat.	Close To
21st	Oph	17h 35m	-24° 28'	-3.9	11"	91%	27° E	PM	*	Jupiter
23rd	Sgr	17h 46m	-24° 37'	-3.9	11"	90%	28° E	PM	*	Jupiter
25th	Sgr	17h 57m	-24° 43'	-3.9	11"	90%	28° E	PM	*	Jupiter
27th	Sgr	18h 8m	-24° 47'	-3.9	11"	89%	29° E	PM	*	Jupiter
29th	Sgr	18h 19m	-24° 47'	-3.9	12"	89%	30° E	PM	**	Moon, Jupiter

Mars and the Outer Planets

Mars
25th

Jupiter
25th

Saturn
25th

Mars

Date	Con.	R.A.	Dec.	Mag.	Diam.	Ill.	Elon.	Vis.	Rat.	Close To
21st	Vir	13h 58m	-11° 19'	1.7	4"	98%	27° W	AM	*	Spica
25th	Vir	14h 8m	-12° 15'	1.7	4"	98%	29° W	AM	*	Mercury
30th	Vir	14h 20m	-13° 22'	1.7	4"	98%	31° W	AM	*	

The Outer Planets

Planet	Date	Con.	R.A.	Dec.	Mag.	Diam.	Elon.	Vis.	Rat.	Close To
Jupiter	25th	Sgr	17h 53m	-23° 17'	-1.8	32"	27° E	PM	*	Venus
Saturn	25th	Sgr	19h 15m	-22° 12'	0.6	16"	48° E	PM	**	
Uranus	25th	Ari	2h 5m	12° 12'	5.7	4"	151° E	PM	***	
Neptune	25th	Aqr	23h 9m	-6° 31'	7.9	2"	107° E	PM	***	

Highlights

Date	Time (UT)	Event
23rd	16:21	The waning crescent Moon is north of the bright star Spica. (Virgo, morning sky.)
	N/A	Good opportunity to see Earthshine on the waning crescent Moon. (Morning sky.)
24th	08:42	The waning crescent Moon is north of Mars. (Morning sky.)
	13:57	Venus is 1.4° south of Jupiter. (Evening sky.)
25th	01:52	The waning crescent Moon is north of Mercury. (Morning sky.)
26th	15:06	New Moon. (Not visible.)
27th	17:38	Neptune is stationary prior to resuming prograde motion. (Evening sky.)
28th	09:51	The waxing crescent Moon is north of Jupiter. (Evening sky.)
	10:22	Mercury is at greatest western elongation from the Sun. (Morning sky.)
	20:12	The waxing crescent Moon is north of Venus. (Evening sky.)
29th	22:15	The waxing crescent Moon is south of Saturn. (Evening sky.)
30th	03:14	The waxing crescent Moon is south of the dwarf planet Pluto. (Evening sky.)
	N/A	Good opportunity to see Earthshine on the waxing crescent Moon. (Evening sky.)

Planet Locations – November 25th

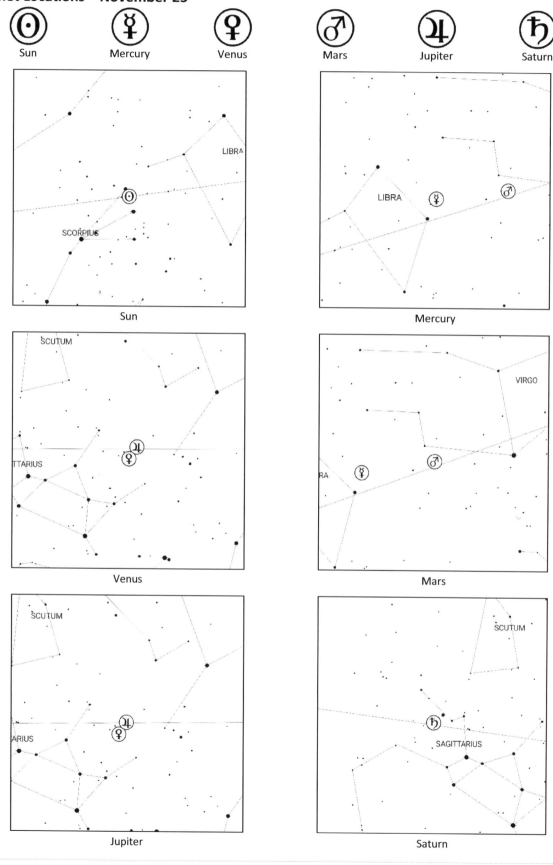

December 1ˢᵗ to 10ᵗʰ, 2019

The Moon

| 1ˢᵗ | 3ʳᵈ | 5ᵗʰ | 7ᵗʰ | 9ᵗʰ |

Date	Con	R.A.	Dec	Mag	Diam	Ill.	Elon.	Phase	Close To
1st	Cap	20h 45m	-20° 43'	-8.7	30'	24%	64° E	+Cr	
2nd	Cap	21h 35m	-17° 59'	-9.3	30'	33%	76° E	+Cr	
3rd	Aqr	22h 23m	-14° 33'	-9.8	30'	43%	86° E	FQ	Neptune
4th	Aqr	23h 9m	-10° 34'	-10.2	30'	52%	97° E	FQ	Neptune
5th	Aqr	23h 54m	-6° 13'	-10.6	30'	61%	107° E	FQ	Neptune
6th	Cet	0h 37m	-1° 36'	-10.9	30'	70%	117° E	+G	
7th	Psc	1h 21m	3° 6'	-11.3	30'	79%	127° E	+G	Uranus
8th	Psc	2h 7m	7° 46'	-11.6	30'	86%	137° E	+G	Uranus
9th	Ari	2h 54m	12° 13'	-11.9	30'	92%	148° E	+G	
10th	Tau	3h 43m	16° 14'	-12.2	31'	97%	159° E	FM	Pleiades, Hyades

Mercury and Venus

Mercury
5ᵗʰ

Venus
5ᵗʰ

Mercury

Date	Con.	R.A.	Dec.	Mag.	Diam.	Ill.	Elon.	Vis.	Rat.	Close To
1st	Lib	15h 10m	-15° 30'	-0.6	6"	70%	20° W	AM	***	
3rd	Lib	15h 19m	-16° 20'	-0.6	6"	75%	19° W	AM	***	
5th	Lib	15h 30m	-17° 12'	-0.6	6"	79%	19° W	AM	***	
7th	Lib	15h 41m	-18° 5'	-0.6	6"	83%	18° W	AM	**	
9th	Lib	15h 52m	-18° 56'	-0.6	5"	86%	18° W	AM	**	Antares

Venus

Date	Con.	R.A.	Dec.	Mag.	Diam.	Ill.	Elon.	Vis.	Rat.	Close To
1st	Sgr	18h 29m	-24° 44'	-3.9	12"	89%	30° E	PM	**	Jupiter
3rd	Sgr	18h 40m	-24° 39'	-3.9	12"	88%	31° E	PM	**	Saturn
5th	Sgr	18h 51m	-24° 30'	-3.9	12"	88%	31° E	PM	**	Saturn
7th	Sgr	19h 2m	-24° 18'	-3.9	12"	88%	32° E	PM	**	Saturn
9th	Sgr	19h 13m	-24° 4'	-4.0	12"	87%	32° E	PM	**	Saturn

Mars and the Outer Planets

Mars
5th

Jupiter
5th

Saturn
5th

Mars

Date	Con.	R.A.	Dec.	Mag.	Diam.	Ill.	Elon.	Vis.	Rat.	Close To
1st	Lib	14h 23m	-13° 36'	1.7	4"	97%	31° W	AM	*	
5th	Lib	14h 33m	-14° 28'	1.7	4"	97%	33° W	AM	*	
10th	Lib	14h 46m	-15° 31'	1.7	4"	97%	35° W	AM	*	

The Outer Planets

Planet	Date	Con.	R.A.	Dec.	Mag.	Diam.	Elon.	Vis.	Rat.	Close To
Jupiter	5th	Sgr	18h 2m	-23° 18'	-1.8	32"	19° E	PM	*	
Saturn	5th	Sgr	19h 19m	-22° 5'	0.6	15"	38° E	PM	**	Venus
Uranus	5th	Ari	2h 4m	12° 6'	5.7	4"	140° E	PM	***	
Neptune	5th	Aqr	23h 9m	-6° 30'	7.9	2"	96° E	PM	***	Moon

Highlights

Date	Time (UT)	Event
4th	06:59	First Quarter Moon. (Evening sky.)
	10:54	The first quarter Moon is south of Neptune. (Evening sky.)
8th	10:08	The waxing gibbous Moon is south of Uranus. (Evening sky.)
9th	N/A	The Monocerotid meteor shower is at its peak. (ZHR: 2)
10th	12:45	The waxing gibbous Moon is south of the Pleiades star cluster. (Evening sky.)

Planet Locations – December 5th

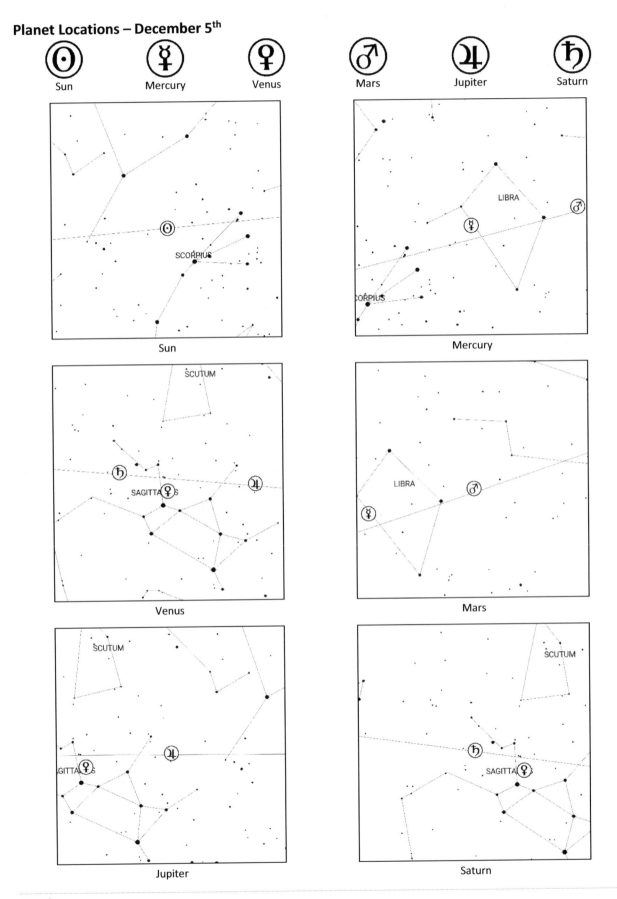

The Moon

| | 11th | | 13th | | 15th | | 17th | | 19th |

Date	Con	R.A.	Dec	Mag	Diam	Ill.	Elon.	Phase	Close To
11th	Tau	4h 36m	19° 34'	-12.5	31'	99%	171° E	FM	Hyades, Aldebaran
12th	Tau	5h 32m	21° 57'	-12.6	31'	100%	176° W	FM	
13th	Gem	6h 30m	23° 8'	-12.3	32'	98%	163° W	FM	
14th	Gem	7h 30m	22° 57'	-12.0	32'	94%	149° W	-G	
15th	Cnc	8h 30m	21° 22'	-11.6	32'	87%	135° W	-G	Praesepe
16th	Leo	9h 28m	18° 28'	-11.3	32'	79%	121° W	-G	Regulus
17th	Leo	10h 25m	14° 29'	-10.9	32'	69%	108° W	-G	Regulus
18th	Leo	11h 19m	9° 40'	-10.5	32'	58%	96° W	LQ	
19th	Vir	12h 12m	4° 21'	-10.0	32'	47%	84° W	LQ	
20th	Vir	13h 4m	-1° 12'	-9.4	32'	35%	72° W	-Cr	Spica

Mercury and Venus

Mercury
15th

Venus
15th

Mercury

Date	Con.	R.A.	Dec.	Mag.	Diam.	Ill.	Elon.	Vis.	Rat.	Close To
11th	Sco	16h 4m	-19° 45'	-0.6	5"	88%	17° W	AM	**	Antares
13th	Sco	16h 17m	-20° 32'	-0.6	5"	90%	16° W	AM	**	Antares
15th	Oph	16h 29m	-21° 16'	-0.6	5"	92%	15° W	AM	**	Antares
17th	Oph	16h 42m	-21° 57'	-0.6	5"	93%	14° W	NV	N/A	Antares
19th	Oph	16h 55m	-22° 33'	-0.6	5"	94%	13° W	NV	N/A	Antares

Venus

Date	Con.	R.A.	Dec.	Mag.	Diam.	Ill.	Elon.	Vis.	Rat.	Close To
11th	Sgr	19h 24m	-23° 46'	-4.0	12"	87%	33° E	PM	**	Saturn
13th	Sgr	19h 34m	-23° 26'	-4.0	12"	86%	33° E	PM	**	Saturn
15th	Sgr	19h 45m	-23° 3'	-4.0	12"	86%	34° E	PM	**	Saturn
17th	Sgr	19h 55m	-22° 37'	-4.0	12"	85%	34° E	PM	**	Saturn
19th	Sgr	20h 6m	-22° 8'	-4.0	12"	85%	35° E	PM	**	

Mars and the Outer Planets

Mars
15th

Jupiter
15th

Saturn
15th

Mars

Date	Con.	R.A.	Dec.	Mag.	Diam.	Ill.	Elon.	Vis.	Rat.	Close To
11th	Lib	14h 49m	-15° 43'	1.7	4"	97%	36° W	AM	*	
15th	Lib	15h 0m	-16° 31'	1.7	4"	97%	38° W	AM	*	
20th	Lib	15h 13m	-17° 28'	1.6	4"	96%	40° W	AM	*	

The Outer Planets

Planet	Date	Con.	R.A.	Dec.	Mag.	Diam.	Elon.	Vis.	Rat.	Close To
Jupiter	15th	Sgr	18h 12m	-23° 17'	-1.8	32"	11° E	NV	N/A	
Saturn	15th	Sgr	19h 24m	-21° 57'	0.6	15"	29° E	PM	**	Venus
Uranus	15th	Ari	2h 3m	12° 1'	5.7	4"	128° E	PM	***	
Neptune	15th	Aqr	23h 10m	-6° 28'	7.9	2"	85° E	PM	***	

Highlights

Date	Time (UT)	Event
11th	04:34	Venus is 1.8° south of Saturn. (Evening sky.)
	11:35	The nearly full Moon is north of the bright star Aldebaran. (Taurus, evening sky.)
12th	05:13	Full Moon. (Visible all night.)
	N/A	The Sigma Hydrid meteor shower is at its maximum. (ZHR: 3)
14th	N/A	The Geminid meteor shower is at its maximum. (ZHR: 120)
15th	11:26	Mercury is 5.2° north of the bright star Antares. (Scorpius, morning sky.)
	15:37	The waning gibbous Moon is north of the Praesepe star cluster. (Cancer, morning sky.)
16th	N/A	The Comae Berenicid meteor shower is at its maximum. (ZHR: 3)
17th	04:43	The waning gibbous Moon is north of the bright star Regulus. (Leo, morning sky.)
19th	04:58	Last Quarter Moon (Morning sky.)
20th	21:05	The just-past last quarter Moon is north of the bright star Spica. (Virgo, morning sky.)
	N/A	The Leonis Minorid meteor shower is at its maximum. (ZHR: 5)

Planet Locations – December 15th

Sun · Mercury · Venus · Mars · Jupiter · Saturn

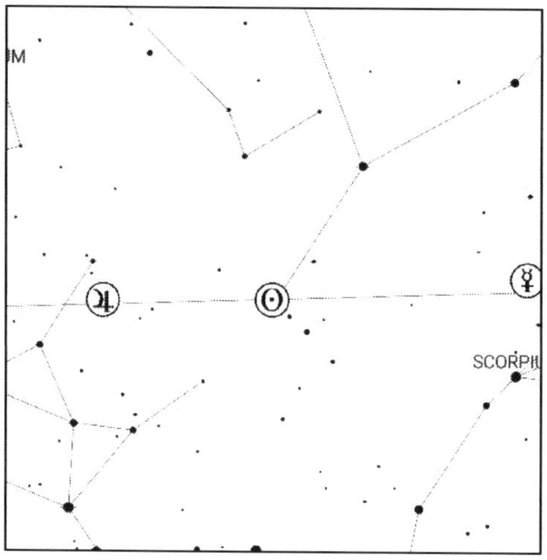

Sun

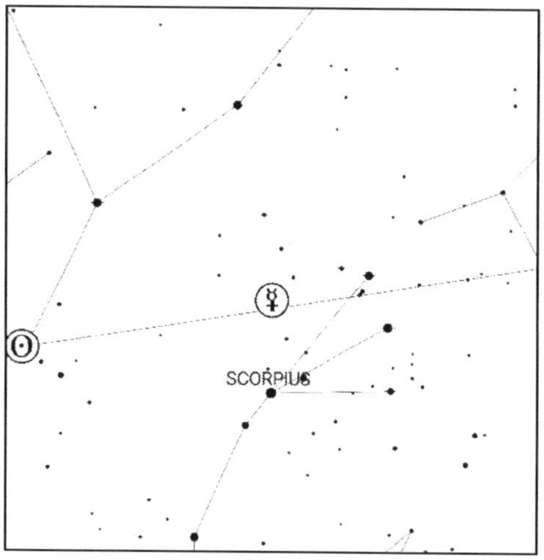

Mercury

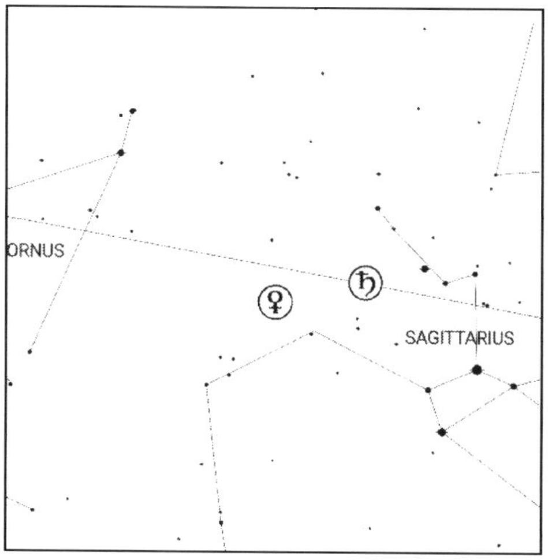

Venus

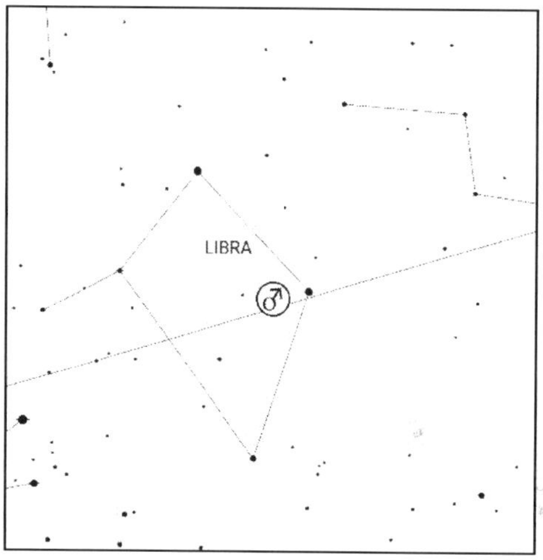

Mars

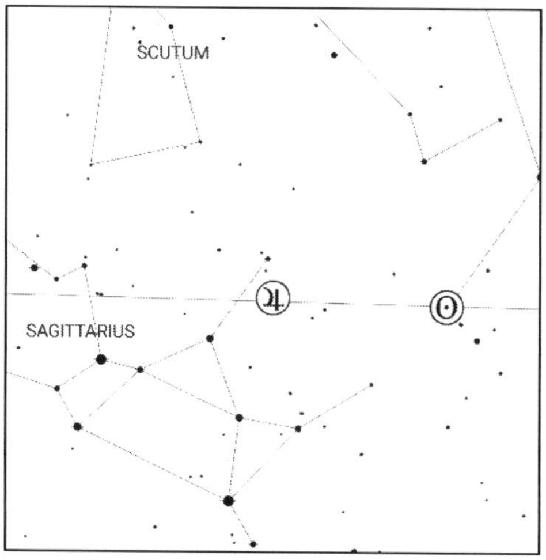

Jupiter

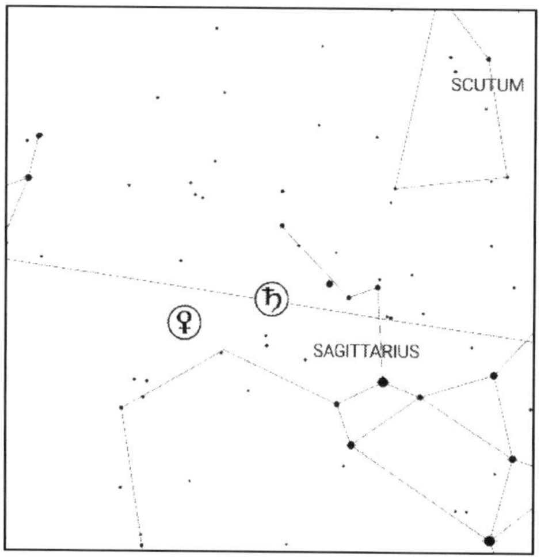

Saturn

December 21ˢᵗ – 31ˢᵗ

The Moon

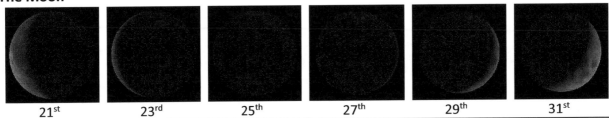

| 21ˢᵗ | 23ʳᵈ | 25ᵗʰ | 27ᵗʰ | 29ᵗʰ | 31ˢᵗ |

Date	Con	R.A.	Dec	Mag	Diam	Ill.	Elon.	Phase	Close To
21st	Vir	13h 56m	-6° 40'	-8.8	32'	25%	60° W	-Cr	Spica
22nd	Lib	14h 49m	-11° 45'	-8.0	32'	16%	48° W	-Cr	Mars
23rd	Lib	15h 43m	-16° 11'	-7.1	32'	9%	36° W	NM	Mars, Antares
24th	Oph	16h 39m	-19° 42'	-6.1	32'	3%	23° W	NM	Antares
25th	Oph	17h 36m	-22° 4'	-4.9	31'	1%	10° W	NM	Mercury, Jupiter
26th	Sgr	18h 34m	-23° 9'	-4.3	31'	0%	4° E	NM	Jupiter
27th	Sgr	19h 30m	-22° 57'	-5.5	31'	2%	17° E	NM	Saturn
28th	Cap	20h 25m	-21° 33'	-6.6	30'	6%	29° E	NM	Venus
29th	Cap	21h 17m	-19° 8'	-7.5	30'	11%	41° E	NM	Venus
30th	Aqr	22h 7m	-15° 54'	-8.2	30'	18%	53° E	+Cr	
31st	Aqr	22h 53m	-12° 4'	-8.8	30'	26%	63° E	+Cr	Neptune

Mercury and Venus

Mercury
25ᵗʰ

Venus
25ᵗʰ

Mercury

Date	Con.	R.A.	Dec.	Mag.	Diam.	Ill.	Elon.	Vis.	Rat.	Close To
21st	Oph	17h 8m	-23° 6'	-0.6	5"	95%	12° W	NV	N/A	Antares
23rd	Oph	17h 21m	-23° 34'	-0.7	5"	96%	11° W	NV	N/A	
25th	Oph	17h 34m	-23° 57'	-0.7	5"	97%	10° W	NV	N/A	Moon
27th	Sgr	17h 48m	-24° 15'	-0.7	5"	98%	9° W	NV	N/A	Jupiter
29th	Sgr	18h 2m	-24° 29'	-0.8	5"	98%	8° W	NV	N/A	Jupiter
31st	Sgr	18h 15m	-24° 37'	-0.8	5"	99%	6° W	NV	N/A	Jupiter

Venus

Date	Con.	R.A.	Dec.	Mag.	Diam.	Ill.	Elon.	Vis.	Rat.	Close To
21st	Cap	20h 16m	-21° 37'	-4.0	12"	85%	35° E	PM	**	
23rd	Cap	20h 27m	-21° 4'	-4.0	13"	84%	35° E	PM	**	
25th	Cap	20h 37m	-20° 28'	-4.0	13"	84%	36° E	PM	**	
27th	Cap	20h 47m	-19° 50'	-4.0	13"	83%	36° E	PM	**	
29th	Cap	20h 57m	-19° 9'	-4.0	13"	83%	36° E	PM	**	Moon
31st	Cap	21h 7m	-18° 27'	-4.0	13"	82%	36° E	PM	**	

Mars and the Outer Planets

Mars
25th

Jupiter
25th

Saturn
25th

Mars

Date	Con.	R.A.	Dec.	Mag.	Diam.	Ill.	Elon.	Vis.	Rat.	Close To
21st	Lib	15h 16m	-17° 39'	1.6	4"	96%	40° W	AM	*	
25th	Lib	15h 27m	-18° 22'	1.6	4"	96%	42° W	AM	*	
31st	Lib	15h 43m	-19° 22'	1.6	4"	96%	44° W	AM	*	

The Outer Planets

Planet	Date	Con.	R.A.	Dec.	Mag.	Diam.	Elon.	Vis.	Rat.	Close To
Jupiter	25th	Sgr	18h 22m	-23° 14'	-1.8	32"	2° E	NV	N/A	Moon
Saturn	25th	Sgr	19h 29m	-21° 48'	0.6	15"	19° E	PM	**	
Uranus	25th	Ari	2h 2m	11° 58'	5.7	4"	117° E	PM	***	
Neptune	25th	Aqr	23h 10m	-6° 25'	7.9	2"	74° E	PM	**	

Highlights

Date	Time (UT)	Event
22nd	04:20	Winter Solstice.
23rd	00:50	The waning crescent Moon is north of Mars. (Morning sky.)
	N/A	The Ursid meteor shower is at its maximum. (ZHR: 10)
	N/A	Good opportunity to see Earthshine on the waning crescent Moon. (Morning sky.)
24th	06:45	The waning crescent Moon is north of the bright star Antares. (Scorpius, morning sky.)
26th	05:14	New Moon. (Not visible.)
	05:19	Annular Solar Eclipse. Visible from far eastern Africa, south Asia, the Indian Ocean and the Pacific.
27th	11:15	The waxing crescent Moon is south of Saturn. (Evening sky.)
	18:25	Jupiter is in conjunction with the Sun. (Not visible.)
29th	01:49	The waxing crescent Moon is south of Venus. (Evening sky.)
	N/A	Good opportunity to see Earthshine on the waxing crescent Moon. (Evening sky.)
31st	22:27	The waxing crescent Moon is south of Neptune. (Evening sky.)

Planet Locations – December 25th

Sun Mercury Venus Mars Jupiter Saturn

Sun

Mercury

Venus

Mars

Jupiter

Saturn

Glossary

Aphelion

The point at which an object is farthest from the Sun. (See also *perihelion*.)

Apogee

The point at which an object is farthest from the Earth. (See also *perigee.*)

Apparent Diameter

Apparent Diameter is the size an object appears in the sky and is measured in degrees, arc-minutes and arc-seconds. If you were to stand facing due north and slowly turn toward east, south, west and then north again, you would be turning 360° (degrees.)

You might therefore think that you can see 360° of sky overhead – in fact, you can only see 180° because the sky is only the visible half a sphere and since you can't see the entire sphere (the ground is in the way,) you can't look 360° in every direction at the sky.

If, however, you were to face due north and look directly overhead at the zenith, this would be 90°. Look down from the zenith to the southern horizon and you would see another 90°, making 180° total. (Incidentally, how high an object appears in the sky is called its *altitude*, but it isn't necessary to know that to use this book.)

To put this into perspective, the Sun and Moon both appear to be about half a degree in diameter but because these are the largest astronomical objects in the sky and everything else appears to be smaller, we need a more convenient (and accurate) measurement.

A degree then is broken up into sixty arc-minutes. Therefore, because the Sun and Moon both appear to be about half a degree, we say their apparent diameter is 30' (arc-minutes.)

The planets and asteroids are even smaller, so we break each arc-minute up into sixty again, thereby creating arc-seconds. A planets' apparent diameter will greatly depend on its actual size and its distance from Earth.

For example, although the planet Venus is slightly smaller than the Earth, it is also the closest world to our own. So at its closest (inferior conjunction) it can have an apparent diameter of 66.01" – in other words, 1' 01" (one arc-minute and one arc-second.) Despite Jupiter being large enough to swallow all the other planets within it, it is much further away – therefore, at its best, it is only able to reach 50.12" (fifty arc-seconds). Neptune is the fourth largest planet but is also the most distant and barely manages to reach 2.37" (arc-seconds.)

Even through a small telescope observers can easily see all of the planets as discs; however, how large the planet appears and the details seen will vary greatly, depending upon the equipment used and the apparent diameter of the planet itself. The dark bands of Jupiter's' atmosphere are easily visible in almost any sized 'scope, whereas Uranus and Neptune typically only show tiny discs in a small to medium sized telescope. Under low power those distant worlds can easily be mistaken for stars.

Only the apparent diameters of the Moon and planets are noted in this book; the dwarf planets Pluto and Ceres, along with the asteroids Juno, Pallas and Vesta and all the bright stars mentioned, only appear as points of light through amateur instruments and their apparent diameter is therefore negligible.

To help put this into perspective, the image below depicts the average apparent size of the planets in comparison with one another.

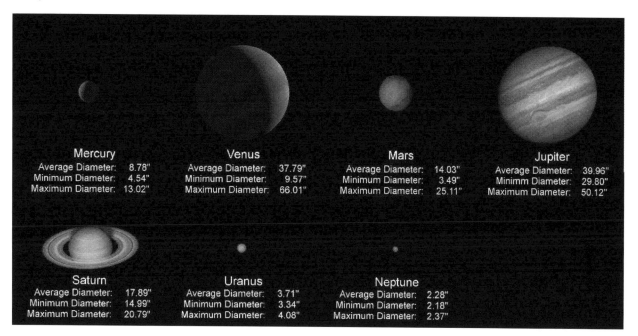

Asterism

An asterism is a recognizable pattern of stars within a constellation. For example, the "backwards question mark" depicting the head of Leo the Lion, or the seven stars of the Big Dipper (aka, the Plough) in the much-larger constellation of Ursa Major, the Great Bear.

AU – Astronomical Unit

An astronomical unit is basically the mean distance of the Earth to the Sun and is the standard measurement of distance within the solar system. One astronomical unit is equivalent to almost 150 million kilometers (specifically, 149,597,870 km) or roughly 92.96 million miles.

Conjunction

Astronomically, this is a fairly vague term. It basically means any situation when two bodies appear close to one another in the sky. However, there is no officially recognized separation limit that would clearly define when a conjunction is taking place. An object in conjunction with the Sun is never visible because the light from the Sun is too over-powering and the object will be lost in the glare. (See also *inferior conjunction, opposition* and *superior conjunction.*)

Culminate

Culmination occurs when an object is at its highest point in the sky. For most objects, this means it is due south, but this will depend on the object and your latitude. For example, some stars or objects may be directly overhead when they culminate. (In fact, if you faced due south and followed an invisible line between due south and due north, your gaze would pass overhead and any object culminating could also be in sight.)

Earthshine

Earthshine is when the Moon is a crescent but you can see the "dark side" of the Moon too – so you can see the whole Moon in the sky. This happens when light is reflected from the daylight side of the Earth and illuminates the unlit portion of the Moons' surface. It can make for a beautiful sight in the twilight, especially when the Moon is close to a bright star or planet.

Left: An example of Earthshine on the waxing crescent Moon. Light is reflected from the Earth, causing the "dark side" of the Moon to be visible. In this image, the unlit portion of the Moon has been lightened to be more apparent. Image by Steve Jurvetson and used under the Creative Commons Attribution 2.0 Generic license.

Ecliptic

The approximate path the Sun, Moon and planets appear to follow across the sky. The ecliptic crosses the traditional twelve signs of the zodiac as well as briefly passing through other constellations, such as Ophiuchus and Orion. The ecliptic is depicted as a pale blue line in the images used throughout this book.

Elongation

Elongation is how far to the east or west an object appears in relation to another object (most usually in relation to the Sun.) If Mercury or Venus is at eastern elongation, it will appear in the evening sky. If Mercury or Venus is at western elongation, it will appear in the pre-dawn sky. (It is worth noting that there is no guarantee the planet will be visible – it will also depend upon the time of year and the observers' latitude.)

Gibbous

The Moon is said to be gibbous between the half phases (first and last quarter) and full Moon. It's hard to describe the shape – it's not a half Moon, but it's not completely circular either. The inner planets Mercury and Venus can also show a gibbous phase. (See also *illumination.*)

Globular Star Cluster

A globular star cluster is, quite literally and simply, a sphere of thousands of stars. Globular clusters appear as faint, misty balls of grey light against the night, and although they all require at least a pair of binoculars to be seen, many can be resolved into their individual stars through a telescope. The best (and most famous) example in the northern hemisphere is M13, the Great Hercules Cluster (see image below and also *open star cluster.*)

Left: M13, the Great Hercules Cluster. Photo taken by the author using Slooh.

Illumination

Simply how much of an object's visible surface is lit. For example, when the Moon is new, none of the lit surface is visible, so the Moon is 0% illuminated. At half phase (first or last quarter) half the lit surface is visible, so the Moon is 50% illuminated. At full Moon, the entire lit surface is visible, so the Moon is 100% illuminated. The planets Mercury, Venus and Mars can also show phases (Mars, being an outer planet, is more limited) and so their illumination will also change over time. (See also *gibbous.*)

Inferior Conjunction

Inferior conjunction occurs when either Mercury or Venus are directly between the Earth and the Sun. Because the planet will appear so close to the Sun in the sky, it will not be visible from Earth. Mercury and Venus are the only two planets that can go through inferior conjunction because only these two worlds orbit closer to the Sun than the Earth. (See also *conjunction* and *superior conjunction.*)

Magnitude

An object's magnitude is simply a measurement of its brightness. The ancient Greeks created a system where the brightest stars were given a magnitude of 1 and the faintest were magnitude 6. Since that time, astronomers have refined the system and increased its accuracy, but as a result, the magnitude range has increased dramatically.

For example, there are some objects that are brighter than zero and therefore have a negative magnitude. Sirius, the brightest star in the sky, has a magnitude of -1.47. All of the naked eye planets – Mercury, Venus, Mars, Jupiter and Saturn – can all have negative magnitudes. The other two planets, Uranus and Neptune, dwarf planets and asteroids are all more than magnitude five.

(The bright star Vega, in the constellation Lyra, is used as the standard reference point. It has a magnitude of 0.0.)

The naked eye can, theoretically, see objects up to magnitude six, but this greatly depends upon the observer's vision and the conditions of the night sky. Most people can see up to around magnitude five under clear, dark, rural skies. Light pollution is so bad in many towns and cities that, even in the suburbs, it is often difficult to see anything fainter than magnitude 3 or 4 at best.

However, the Moon, Mercury, Venus, Mars, Jupiter and Saturn should easily be visible to anyone, anywhere, assuming that the object is not too close to the Sun. A planets' magnitude will vary, depending upon how close it is to the Earth, how large it appears in the sky and how much of its lit surface is visible. (See also *apparent diameter* and *illumination*.)

Constellations can be problematic, depending upon the brightness of the stars that form the constellation itself.

Meteors are best observed from rural skies but the bright stars mentioned, as well as the Pleiades and Hyades star clusters can be seen from the suburbs.

Almost everything else is best observed under rural skies and/or with binoculars or a telescope.

The magnitude ranges of the solar system objects mentioned in this book are detailed below. (The Sun is, on average, about magnitude -26.74)

Object	Minimum Magnitude (Faintest)	Maximum Magnitude (Brightest)
Moon	-2.5 (New)	-12.9 (Full)
Mercury	5.73	-2.45
Venus	-3.82	-4.89
Mars	1.84	-2.91
Jupiter	-1.61	-2.94
Saturn	1.47	-0.49
Uranus	5.95	5.32
Neptune	8.02	7.78
Pluto	16.3	13.65
Dwarf planet Ceres	9.34	6.64
Asteroid 2 Pallas	10.65	6.49
Asteroid 3 Juno	11.55	7.4
Asteroid 4 Vesta	8.48	5.1

Occultation

When one object completely covers another. Many occultations involve the Moon occulting a star or, sometimes, a planet, but on occasion, a planet may be seen to occult a star. On very rare occasions, one planet may occult another.

Open Star Cluster

An open star cluster is one where the stars appear to be loosely scattered against the night. Unlike globular clusters, their member stars are usually quite young and only number a couple of hundred at most. Several open clusters can be seen with the naked eye (for example, M44 - the Praesepe - in Cancer and the Hyades, which forms the V shaped asterism in the constellation Taurus.) The most famous example of an open cluster is M45, the Pleiades, a naked eye cluster (also in Taurus) that is easily seen throughout the winter. (See image below and also *globular star cluster*.)

Left: M45, the famous Pleiades open star cluster in Taurus. Easily visible with the naked eye throughout the winter, this image shows the deep blue nebulosity that surround the stars. This nebulosity is the remains of the cloud that gave birth to the stars themselves, but unfortunately this is not visible to the vast majority of observers. Photo by the author using Slooh.

Opposition

An object is said to be at opposition when it is directly opposite the Sun in the sky. On that date, it is visible throughout the night as it will rise at sunset, culminate at midnight and set at sunrise. For that reason, this is the best opportunity to observe that object. (See also *conjunction*.)

Perihelion

The point at which an object is closest to the Sun. (See also *aphelion*.)

Perigee

The point at which an object is closest to the Earth. (See also *apogee*.)

Prograde Motion

Prograde motion is when a body appears to move forwards through the constellations from west to east. It's the normal motion of the Sun, Moon, planets and asteroids across the sky. (See also *retrograde motion.*)

Retrograde Motion

Retrograde motion is when a body appears to move *backwards* through the constellations, from east to west. For the inferior planets Mercury and Venus, this happens for a time after greatest eastern elongation (when the planet appears in the evening sky) and before greatest western elongation (when it appears in the pre-dawn sky) as the planet catches up to and then passes the Earth in its orbit. For all the other planets and asteroids, it happens for a time before and after opposition when the Earth catches up to that world and then passes it. This YouTube video from 2009 does a good job of graphically depicting how this happens. (See also *prograde motion.*)

Superior Conjunction

Superior conjunction occurs when either Mercury or Venus are on the opposite side of the Sun from the Earth. For example, if Mercury is at superior conjunction, the Sun would be directly between the Earth and the Mercury. Like *inferior conjunction*, the planet appears very close to the Sun in the sky and is not visible from Earth.

Again, like *inferior conjunction,* Mercury and Venus are the only two planets that can go through inferior conjunction because only these two worlds orbit closer to the Sun than the Earth. (See also *conjunction* and *inferior conjunction.*)

Universal Time

Universal Time is the standard method of notating when an astronomical event takes place. It is based upon Greenwich Mean Time and requires adjustment for other time zones:

Greenwich Mean Time – no change. (Summer Time – add one hour.)

Eastern Time – deduct five hours. (Summer Time – deduct four hours.)

Central Time – deduct six hours. (Summer Time – deduct five hours.)

Mountain Time – deduct seven hours. (Summer Time – deduct six hours.)

Pacific Time – deduct eight hours. (Summer Time – deduct seven hours.)

A useful website that will convert Universal Time to other time zones can be found at the following address: http://www.worldtimeserver.com/convert_time_in_UTC.aspx

Bear in mind that if an event takes place during the day at your location, it may still be visible in the evening or pre-dawn sky. For example, two planets may be at their closest at 1pm local time but because they don't move quickly, they'll still be very close together during the night. The only exception is the Moon – it *does* move relatively quickly, but may still appear fairly close to the object when it next becomes visible. It just won't be as close as it was at the time listed in the book.

Waning

The Moon is said to be "waning" between the full and new Moon. When the Moon is full, the Earth lies between the Moon and the Sun and the lit surface is completely visible to us. When the Moon wanes, the visible, lit portion of the Moon appears to decrease until it is completely invisible at new Moon. A waning Moon is best seen in the pre-dawn sky. (See also *waxing*.)

Waxing

The Moon is said to be "waxing" between the new and full Moon. When the Moon is new, it lies between the Earth and the Sun and the lit surface is not visible to us. When the Moon waxes, the visible, lit portion of the Moon appears to be increasing until it is completely lit at full Moon. A waxing Moon is best seen in the evening sky. (See also *waning*.)

Zenith

The point directly overhead in the sky. (See also *zenith hourly rate.*)

Zenith Hourly Rate

The number of meteors an observer can expect to see each hour at the zenith (directly overhead) on the shower's peak date. It's worth remembering that meteor showers can be somewhat unpredictable and, hence, the maximum zenith hourly rate is only an estimate at best. (See also *zenith.*)

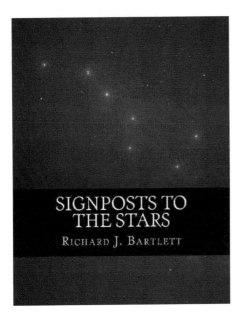

Signposts to the Stars

Easy Things to See With a Small Telescope

Aimed at absolute beginners, this book will help you to locate and learn the constellations using the brightest stars of Ursa Major and Orion as signposts.

More than that, the book also details:
*Key astronomical terms and phrases
*The brightest stars and constellations for each season
*The myths and legends of the stars
*Fascinating stars, star clusters, nebulae and galaxies, many of which can be seen with just your eyes or binoculars
*An introduction to the planets, comets and meteor showers

If you've ever stopped and stared at the stars but didn't know where to begin, these signposts will get you started on your journey!

Specifically written with the beginner in mind, this book highlights over sixty objects easily found and observed in the night sky. Objects such as:
* Stunning multiple stars
* Star clusters
* Nebulae
* And the Andromeda Galaxy!

Each object has its own page which includes a map, a view of the area through your finderscope and a depiction of the object through the eyepiece.

There's also a realistic description of every object based upon the author's own notes written over years of observations.

Additionally, there are useful tips and tricks designed to make your start in astronomy easier and pages to record your observations.

If you're new to astronomy and own a small telescope, this book is an invaluable introduction to the night sky.

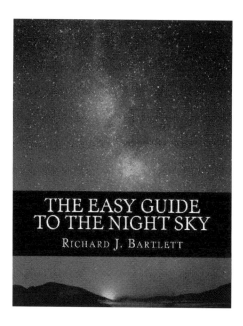

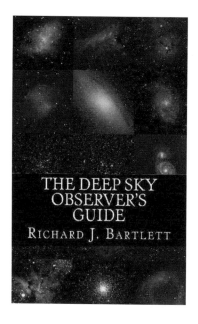

The Easy Guide to the Night Sky

The Deep Sky Observer's Guide

Written for the amateur astronomer who wants to discover more in the night sky, this book explores the constellations and reveals many of the highlights visible with just your eyes or binoculars.

The highlights include:
* The myths and legends associated with the stars
* Bright stars and multiple stars
* Star clusters
* Nebulae
* Galaxies

Each constellation has its own star chart and almost all are accompanied by graphics depicting the highlights and binocular views of the best objects.

Whether you're new to astronomy or are an experienced stargazer simply looking to learn more about the constellations, this book is an invaluable guide to the night sky and the stars to be found there.

The Deep Sky Observer's Guide offers you the night sky at your fingertips. As an amateur astronomer, you want to know what's up tonight and you don't always have the time to plan ahead. The Deep Sky Observer's Guide can solve this problem in a conveniently sized paperback that easily fits in your back pocket. Take it outside and let the guide suggest any one of over 1,300 deep sky objects, all visible with a small telescope and many accessible via binoculars.

* Multiple stars with 2" or more of separation
* Open clusters up to magnitude 9
* Nebulae up to magnitude 10
* Globular clusters up to magnitude 10
* Planetary nebulae up to magnitude 12
* Galaxies up to magnitude 12
* Includes lists of deep sky objects for the entire sky with R.A. and declination for each and accompanying images for many

Whether you use a GoTo or prefer to star hop, no matter where you live in the world and no matter what time of year or night, the Deep Sky Observer's Guide is the indispensable companion for every adventure among the stars.

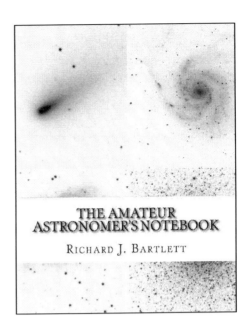

The Wonder of it All

The Amateur Astronomer's Notebook

From our home here on Earth, past the Sun, Moon and planets, this is a journey out to the stars and beyond.

A journey of discovery that shows us the beauty and wonder of the cosmos and our special and unique place within it.

Written by an amateur astronomer with a life-long love of the stars, The Wonder of It All will open your child's eyes to the universe and includes notes for parents to help develop an interest in astronomy.

The Amateur Astronomer's Notebook is the perfect way to log your observations of the Moon, stars, planets and deep sky objects.

With an additional appendix with hundreds of suggested deep sky objects, this 8.5" by 11" notebook allows you to record everything you need for 150 observing sessions under the stars:
*Date
*Time
*Lunar Phase
*Limiting Magnitude
*Transparency
*Seeing
*Equipment
*Eyepieces
*Additional Notes
*Pre-drawn circles to sketch your observations
*Plenty of room to record your notes and impressions

Whether you're an experienced astronomer or just beginning to discover the universe around us, you'll find the notebook to be an invaluable tool and record of your exploration of the cosmos.

Printed in Great Britain
by Amazon